Rupert Deakin

Rider Papers on Euclid

Books I and II. Graduated and arranged in order of difficulty, with an introduction on teaching Euclid

Rupert Deakin

Rider Papers on Euclid

Books I and II. Graduated and arranged in order of difficulty, with an introduction on teaching Euclid

ISBN/EAN: 9783337164171

Printed in Europe, USA, Canada, Australia, Japan

Cover: Foto ©Paul-Georg Meister /pixelio.de

More available books at **www.hansebooks.com**

RIDER PAPERS ON EUCLID

(Books I. and II.)

GRADUATED AND ARRANGED IN ORDER OF DIFFICULTY

WITH AN INTRODUCTION ON TEACHING EUCLID

BY

RUPERT DEAKIN, M.A.
BALLIOL COLLEGE, OXFORD
HEADMASTER OF KING EDWARD'S SCHOOL, STOURBRIDGE

London
MACMILLAN AND CO.
AND NEW YORK
1891

CONTENTS.

	PAGE
Introduction on Teaching Euclid, - - - -	7
Part I. Papers I.-VI. to Euclid I. 12, - - -	11
Part II. Papers VII.-XII. to Euclid I. 26, - -	16
Part III. Papers XIII.-XVIII. to Euclid I. 32, -	21
Part IV. Papers XIX.-XXIV. to Euclid I. 34, -	26
Part V. Papers XXV.-XXX. to Euclid I. 34 (harder),	31
Part VI. Papers XXXI.-XXXVI. to Euclid I. 41, -	36
Part VII. Papers XXXVII.-XLII. to Euclid I. 48, -	41
Part VIII. Papers XLIII.-XLVIII. to Euclid I. 48 (harder), - - - - - - - -	46
Part IX. Papers XLIX.-LIV. on Euclid, Book II., -	52
Propositions in Euclid connected with the Riders, -	57
Enunciations of Propositions in Euclid I. and II., -	59
Examination Papers in Euclid I. and II., - - -	67

INTRODUCTION.

ON TEACHING EUCLID.

This little book has been written specially for my own classes and parts of it have been in use for several years.

In teaching Euclid the first aim should be to get the Definitions, Postulates, Axioms, and Propositions 1 to 12 in Book I. known thoroughly by every boy in the class. Then the Rider Papers in Part I. of this book may be given to be answered. They will be found quite easy enough for boys to answer at home, and if one paper is set each week, Part I. will be sufficient for half a term. My own plan has been to look over each boy's answers and mark them; on the next day to return them to the boys and go through on the blackboard such Riders as have not been answered by the majority of the boys in the class. I have usually found fifteen minutes ample time for this work.

In writing and arranging these Papers I have

constantly kept in view the difficulties that experience shows me all students feel more or less in solving Riders. The first of these difficulties is the inability to draw a proper figure. In the first part of these Papers I have therefore asked for different figures to be drawn; and in all these cases I mean drawn without Proof. Every student should also draw a figure of each Proposition in Euclid, and it is a good plan to draw these figures in an exercise book, one on each page, so that they may be used for saying the Propositions.

Another difficulty to beginners arises from the general terms in which Propositions are usually stated. For example, almost all editions of Euclid contain this Rider :— "The straight line drawn from the vertex of an isosceles triangle to the middle point of the base is perpendicular to the base." Boys who have learnt Euclid for years will refuse to attempt the Rider in this form. But the same Rider may be stated thus:—"Draw an isosceles triangle ABC, having the side AB equal to the side AC. Bisect the base BC in D and join AD. Prove that the angles ADB and ADC are right angles." In this form the Rider will be solved by almost every boy who has learnt the first twelve Propositions in Euclid. Throughout these Papers therefore all Riders, except the simplest, are stated first as Particular Propositions, and afterwards the most important Riders are repeated as General Propositions.

INTRODUCTION.

It would be a great gain to education if we could get rid of the idea that there are a limited number of important Propositions, all contained in Euclid, which must be learnt and remembered; but that there are also an endless number of unimportant Riders, which no one ever can remember. We should rather aim at teaching our pupils that there are different methods of Proof, and that different Propositions or Riders, whether in Euclid or not, are examples of these methods, and serve, just like the examples in Arithmetic or Algebra, to illustrate the different methods of proceeding. It is true that the results we obtain vary in value; but it is also true that many of the most important Propositions are not to be found in Euclid. In teaching Euclid therefore it is a good plan to treat all the Propositions in Book I. as Riders. Before setting a Proposition to be learnt, call the class round the blackboard; state the enunciation, and draw the figure; and then ask anyone to guess how it is proved. In this way the learning of Euclid is made interesting, and the working of Riders is looked upon as the solution of a number of puzzles rather than as an odious task.

The Riders in this book are all important Propositions. The student who has worked through them will be acquainted with all the chief results arrived at in that part of elementary Geometry of which they treat.

The Papers in each Part are graduated in difficulty. They are also often arranged in pairs, so that the solution of any Paper marked with an even number will be found easy after working the preceding Paper.

RUPERT DEAKIN.

King Edward's School,
Stourbridge, *February*, 1891.

RIDER PAPERS.

PART I.

TO EUCLID I. 12.

(*In Papers I. to VI. "Draw" means "Draw without Proof."*)

I.

1. Draw an isosceles triangle having each of the sides double of the base.

2. Draw a right-angled isosceles triangle, an obtuse-angled isosceles triangle and an acute-angled isosceles triangle.

3. Take a straight line MABN divided into three equal parts at A and B. With centre A and radius AN describe the circle NCD. With centre B and radius BM describe the circle MCD, cutting the former circle in the points C and D. Join CA, CB, DA, DB. Prove that BC=AC, and AD=BD. What kind of figure is CADB?

4. In the figure of Question 3 join CD, and prove that the angle ACD is equal to the angle BCD.

5. In the figure of Prop. 2, let the given point A be on the circumference of the smaller circle. Draw

the complete figure in this case. Where does the point D come?

6. Which of the twelve Axioms apply to magnitudes of all kinds, and which apply to geometrical magnitudes only?

II.

1. Explain Proposition, Enunciation, Data and Quæsita.

2. Draw a right-angled scalene triangle, an obtuse-angled scalene triangle and an acute-angled scalene triangle.

3. Take a straight line AB, and produce it both ways to M and N. Make MB=AN. Show how to describe an isosceles triangle on AB, having its two sides CA and CB each equal to MB or AN.

4. ABC is an isosceles triangle, having the side AB equal to the side AC, and the angle BAC is bisected by the straight line AD. Prove that AD also bisects the base BC.

5. In the figure of Prop. 2, let the given point A be joined to C instead of to B. Draw the complete figure in this case.

6. In Prop. 9, why is the equilateral triangle DEF described on the side remote from A? Draw figures to illustrate your answer.

III.

1. Explain Problem, Theorem, Q.E.D. and Q.E.F.

2. There are seven kinds of triangle. Draw one triangle of each kind and give its name.

3. In Prop. 9, prove that AF bisects the angle DFE.

4. ABC is an isosceles triangle, having the side AB equal to the side AC, and the angle BAC is bisected by the straight line AD. Prove that AD is perpendicular to the base BC.

5. The straight line AB is bisected at the point C; and from C the straight line CD is drawn at right angles to AB. In CD take any point E, and join AE and BE. Prove that AE=BE.

6. Write out all the Definitions, Axioms and Postulates that Euclid employs in Prop. 2.

IV.

1. What is a Postulate? Euclid assumes a fourth Postulate in Prop. 4. What is it?

2. Explain, with examples, Reductio ad Absurdum, Converse and Corollary.

3. Draw a quadrilateral figure ABCD, having its opposite sides equal, viz. AB to CD and AD to BC. Join BD. Prove that the angle BAD is equal to the angle BCD.

4. ABC is an isosceles triangle, having AB equal to AC. The angle ABC is bisected by the straight line BD, and the angle ACB by the straight line CD. Prove that DB=DC.

5. In the figure of Prop. 10, take any point E in CA; and from CB cut off CF equal to CE. Join DE and DF. Prove that DE=DF.

6. Show by drawing triangles that two triangles may have all their angles equal, each to each, and yet not be equal in area.

V.

1. What is an Axiom? Why is the twelfth Axiom objectionable?

2. In the figure of Prop. 5, if FC and BG meet in H, prove that HB=HC.

3. ABCD is a rhombus. Join AC. Prove that the angle BAC is equal to the angle DAC.

4. ABC and DBC are two isosceles triangles on the same base BC, on the same side of BC, the vertex A being within the triangle DBC. Join AD. Prove that the angle BDA is equal to the angle CDA.

5. A and B are two given points, and CD is a given line not passing through either A or B. Join AB, and bisect it at E. Through E draw EF at right angles to AB, and meeting CD in F. Join AF and BF. Prove that AF=BF.

6. Take a right angle BAC, and draw a complete figure showing how it may be divided into four equal parts, as in Prop. 9.

VI.

1. Why is the given line of unlimited length in Prop. 12?

2. In the figure of Prop. 5, if FC and BG meet in H, prove that FH=GH.

3. ABC is an isosceles triangle. From the equal sides AB and AC cut off BD equal to CE, and join CD and BE. Prove that CD=BE.

4. Two isosceles triangles ABC and DBC are on the same base BC. Prove that the angle DBA is equal to the angle DCA.

5. Draw two isosceles triangles ABC, ACD, such that AB=AC=AD, and let them have AB and AD in one straight line.

6. In the figure of Question 5, prove that the angle BCD is equal to the sum of the angles ABC and ADC.

PART II.

TO EUCLID I. 26.

VII.

1. Upon a given finite straight line describe an isosceles triangle having each of its equal sides double of the base.

2. In the figure of Prop. 5, join FG, and prove that the angle BGF is equal to the angle CFG.

3. The straight line which bisects the vertical angle of an isosceles triangle also bisects the base, and is perpendicular to it.

4. Let the straight line AB make with the straight line CD the angles ABC and ABD, and let these angles be bisected by the straight lines BE and BF. Prove that EBF is a right angle.

5. ABC is any triangle, and the angle BAC is bisected by the straight line AX which meets BC in X. Prove that BA is greater than BX, and CA is greater than CX.

6. ABC is any triangle. Prove that the difference between any two sides, BA and AC, is less than the third side BC.

VIII.

1. AB is a given straight line, and C and D are two points outside the line AB. Find a point X in AB, such that CX=DX.

2. In the figure of Prop. 5, let FC and BG meet in H, and join AH. Prove that AH bisects the angle BAC.

3. If two isosceles triangles are on the same base, prove that the straight line, produced if necessary, which joins their vertices will bisect their common base, and be perpendicular to it.

4. Two straight lines AB and CD cut each other in the point O, and the four angles at the point O are bisected by the lines OE, OF, OG and OH. Prove that OE and OG, and that OF and OH are in the same straight lines; and that EOG and FOH cut each other at right angles.

5. AB is a straight line, and C a point without it. Draw CD perpendicular to AB, and prove that CD is less than any other line, such as CE, drawn from C to AB.

6. Take any point O inside the triangle ABC, and join OA, OB, OC. Prove that OA, OB and OC are together greater than half the sum of AB, BC and CA.

IX.

1. Show how to draw a straight line any point in which is equidistant from two given points A and B.

2. ABCD is a quadrilateral figure, and its diagonals AC and BD bisect each other at right angles. Prove that ABCD is a rhombus.

3. Show how to divide a given rectilineal angle into two parts so that one part is one-seventh of the other part.

4. A and B are two points in the same side of the line CD. Draw AP perpendicular to CD and produce it to E, making PE equal to AP. Join BE, cutting CD in X; and join AX. Prove that AX and BX make equal angles with CD.

5. ABCD is a quadrilateral figure of which AD is the longest side and BC the shortest. Prove that the angle ABC is greater than the angle ADC, and the angle BCD greater than the angle BAD.

6. In Prop. 16 prove that the two sides AB, BC are together greater than twice the median BE, which bisects the remaining side AC.

X.

1. Show how to find a point equidistant from three given points, which are not in the same straight line.

2. In an isosceles triangle two of the medians are equal.

3. From two given points on opposite sides of a given straight line show how to draw two straight lines which shall meet in the given straight line and make equal angles with it.

4. In any triangle the sum of the medians is less than the perimeter of the triangle.

5. O is any point within the triangle ABC. Prove that OA, OB, OC are together less than AB, BC, and CA together.

6. ABC is any triangle. Through P, the middle point of AB, draw any straight line QPR meeting CB in Q and CA produced in R. From PR cut off PN equal to PQ and join AN. Prove that the triangle APN is equal to the triangle PQB, and that the triangle ABC is less than the triangle QRC.

XI.

1. Show how to find the centre of a circle which shall pass through two given points and have its radius equal to a given line.

2. The three medians of an equilateral triangle are equal.

3. Given two points A and B on the same side of a line CD. Find a point P in CD, such that the sum of AP and BP is a minimum.

4. X, Y, Z are the middle points of the sides BC, CA, AB of the triangle ABC; and YO and ZO are drawn at right angles to CA and AB. Join OX and prove that OX is perpendicular to BC.

5. If a line be divided into any two unequal parts, the distance of the point of section from the middle of the line is equal to half the difference of the two parts.

6. If two right-angled triangles have their hypothenuses equal, and one side of the one equal to one side of the other, the two triangles shall be equal in all respects.

XII.

1. Find a point which is equidistant from four fixed points. When is this impossible?

2. If two circles cut one another, the line joining their points of intersection is bisected at right angles by the line joining their centres.

3. The line AB is drawn at right angles to CD from the middle point of CD. Show how to describe on the base CD an isosceles triangle having the sum of one of the equal sides and the perpendicular drawn from the vertex to the base equal to AB.

4. Any point P is taken on the line AF which bisects the given rectilineal angle BAC, and PX, PY are drawn perpendicular to BA and AC. Prove that PX = PY.

5. If the line AB be bisected at C and produced to D, prove that CD is equal to half the sum of AD and BD.

6. If two triangles have two sides of the one equal to two sides of the other, each to each, and have likewise the angles opposite to one pair of equal sides equal, then the angles opposite to the other pair of equal sides shall be either equal or supplementary, and in the former case the triangles shall be equal in all respects.

PART III.

TO EUCLID I. 32.

XIII.

1. Prove Prop. 8 by supposing the two triangles placed on opposite sides of the same base and their vertices joined.

2. The angles ABC, ACB of the triangle ABC are bisected by the lines BO and CO. Join OA and prove, by drawing perpendiculars from O to the sides of the triangle, that OA bisects the angle BAC.

3. If a straight line falling on two other straight lines makes the two exterior angles on the same side of it together equal to two right angles, these two straight lines shall be parallel.

4. Straight lines which are perpendicular to the same straight line are parallel.

5. ABC is a triangle having the side BA produced to D, and the angle CAD is bisected by the line AX. If AX is parallel to BC, prove that ABC is an isosceles triangle.

6. Each of the angles of an equilateral triangle is two-thirds of a right angle.

XIV.

1. ABCD is a quadrilateral figure, and the two opposite sides AB and CD are bisected at P and Q. Join PQ. If PQ is at right angles to AB and CD, prove that AD = BC.

2. The sides AB and AC of the triangle ABC are produced to D and E, and the angles DBC and ECB are bisected by the lines BO and CO. Prove that AO will bisect the angle BAC.

3. Any straight line parallel to the base of an isosceles triangle makes equal angles with the sides.

4. If a straight line meets two or more parallel straight lines and is perpendicular to one of them, it is perpendicular to all the others.

5. The straight line parallel to the base of an isosceles triangle through the vertex will bisect the exterior angle at the vertex.

6. In a right-angled isosceles triangle each of the equal angles is half a right angle.

XV.

1. From a given point draw a straight line equal to twice a given straight line.

2. If in a quadrilateral two opposite sides be equal, and the angles which a third side makes with the equal sides be equal, then the other angles of the quadrilateral shall be equal.

3. A is any point outside, and B and C any two points in a given straight line. Join AC. With centre

A and radius equal to BC describe a circle ; and with centre B and radius equal to AC describe a circle cutting the former circle in D. Join AD and prove that AD is parallel to BC.

4. If two straight lines AB, AC both pass through the same point A, they cannot both be parallel to another line.

5. AX is a given finite straight line, and P and Q are two given acute angles. Show how to construct a triangle ABC having the angle ABC equal to P, the angle ACB equal to Q, and AX the perpendicular from A to the base BC.

6. If two triangles have two angles of the one equal to two angles of the other, each to each, then the third angle of the one is equal to the third angle of the other.

XVI.

1. In the figure of Prop. 2 show how to draw from the point D a straight line, so that the part of it intercepted between the two circles may be equal to BC.

2. Prove the first case in Prop. 26 by the method of superposition.

3. ABCD is a quadrilateral figure having the side AB equal to the side CD, and the angle ABC to the angle BCD. Prove by superposition that AD is parallel to BC.

4. Through a given point only one straight line can be drawn parallel to a given straight line.

5. Construct a right-angled triangle, having one side equal to a given finite straight line, and one angle equal to a given acute angle.

6. In any right-angled triangle the two acute angles are complementary.

XVII.

1. Prove by the method of superposition that only one perpendicular can be drawn to a given straight line from a given point without it.

2. BAC and BDC are two triangles on the same base BC and on the same side of it, and the angle BAC is equal to the angle BDC. Prove that each of the vertices A and D must lie without the other triangle.

3. Find the locus of a point equidistant from two given intersecting straight lines.

4. Straight lines which make equal angles with two given intersecting straight lines form two sets of parallel lines.

5. Through a given point draw as many straight lines as possible making a given angle with a given straight line.

6. The sum of the angles of any quadrilateral figure is equal to four right angles.

XVIII.

1. From a given point outside a given straight line, not more than two straight lines can be drawn equal to a given straight line, one on each side of the perpendicular from the given point to the given line.

2. ACB and ADB are two triangles on the same base AB and on the same side of it, and AC is equal to BD, and AD to BC. If AD and BC intersect in P, prove that the triangle APB is isosceles.

3. Find a point within a given triangle equidistant from the three sides.

4. Straight lines which make a given acute or obtuse angle with a given straight line form two sets of parallel lines.

5. If one angle of a triangle is equal to the sum of the other two angles, the triangle is right angled.

6. Show how to trisect a right angle.

PART IV.

TO EUCLID I. 34.

XIX.

1. AB, AC are two straight lines. Through the given point X draw a straight line meeting AB and AC in D and E and making AD equal to AE.

2. Construct a triangle having given the base, the altitude and the length of the median which bisects the base.

3. In a right-angled triangle if a perpendicular be drawn from the right angle to the hypothenuse, the two triangles thus formed are equiangular to one another.

4. Every right-angled triangle can be divided into two isosceles triangles by a straight line drawn from the right angle to the hypothenuse, and this line is equal to half the hypothenuse.

5. What is the magnitude of each of the angles of a regular pentagon?

6. If the diagonals of a parallelogram are equal all its angles are right angles.

XX.

1. If in Prop. 33 the lines were joined, but not towards the same parts, state and prove what difference there would be in the conclusion.

2. AB and C are two given straight lines. At the point B the angle ABD is made equal to half a right angle. Find a point P in BD, and a point Q in AB, so that AQP shall be a right-angled triangle having its hypothenuse equal to C, and the sum of its sides equal to AB.

3. A is the vertex of an isosceles triangle. Produce BA to D, making AD equal to AB; and join CD. Prove that BCD is a right angle.

4. A number of right-angled triangles have a common right angle and equal hypothenuses. Show that the middle points of the hypothenuses all lie on the circumference of the same circle.

5. What is the magnitude of each of the angles of a regular hexagon?

6. Two straight lines drawn from the extremity of the base of any triangle cannot bisect each other.

XXI.

1. In the figure of Prop. 16 if the angle ABC is bisected by the line BX meeting AC in X, prove that the median BE falls within the angle ABX so long as AB is greater than BC.

2. If ABC is a triangle having the angles at A and B equal to half two given angles P and Q, and if at the point C the angles ACD, BCE be described equal

respectively to half P and Q, the lines CD, CE meeting the base AB within the triangle, then CDE will be a triangle having its perimeter equal to AB, and the angles at the base equal to P and Q.

3. Draw a straight line at right angles to a given finite straight line from one of its extremities without producing the given straight line.

4. Prove indirectly that if the bisectors of two angles of a triangle are equal the two angles are equal.

5. What is the magnitude of each of the exterior angles of a regular octagon?

6. The straight lines which bisect two opposite angles of a parallelogram are either coincident or parallel.

·XXII.

1. In the figure of Prop. 16 if the angle ABC be bisected by BX, and BP be drawn perpendicular to AC, prove that the bisector BX is intermediate in position and magnitude to the median BE and the perpendicular BP so long as AB and BC are unequal.

2. If ABC is a triangle having the angle at B equal to half a given angle P, and the side AC equal to a given line K, show how to describe on the base AB a triangle having the difference of the base angles equal to P and the difference of the sides equal to K.

3. A parallelogram is bisected by any straight line which passes through the middle point of one of its diagonals.

4. If one angle of a parallelogram is a right angle all its angles are right angles.

5. If the opposite sides of a quadrilateral figure are equal it is a parallelogram.

6. The straight lines which bisect two adjacent angles of a parallelogram intersect at right angles.

XXIII.

1. In any triangle the angle contained by the bisector of the vertical angle and the perpendicular from the vertex to the base is equal to half the difference of the angles at the base of the triangle.

2. ABCD is a quadrilateral figure having AB parallel to CD. Prove that its area is equal to the area of a parallelogram formed by drawing through M the middle point of BC a straight line parallel to AD.

3. From the extremities of the base of the isosceles triangle ABC, BP and CQ are drawn perpendicular to the equal sides AC and AB. Prove that each of the angles PBC and QCB is equal to half the angle BAC.

4. The diagonals of a parallelogram bisect each other.

5. If the opposite angles of a quadrilateral figure are equal it is a parallelogram.

6. The lines which bisect the angles of any parallelogram form a right-angled parallelogram, whose diameters are parallel to the sides of the former parallelogram.

XXIV.

1. On a given base construct a triangle, having one angle equal to a given angle A, and the side opposite this angle equal to a given straight line B. When will

there be two solutions, one solution, or no solution possible?

2. ABC is any triangle and CPQ is drawn perpendicular to the bisector of the angle A meeting it in P, and the side AB in Q. Prove that AQC is an isosceles triangle, and that the angle QCB is equal to half the difference of the angles ABC and ACB.

3. If a quadrilateral figure has all its sides equal and one angle a right angle, all its angles are right angles.

4. If the diagonals of a quadrilateral figure bisect each other, the figure is a parallelogram.

5. If a quadrilateral figure has two of its opposite sides parallel, and the other two sides equal but not parallel, any two of its opposite angles are together equal to two right angles.

6. The parts of all perpendiculars to two parallel lines intercepted between them are equal.

PART V.

TO EUCLID I. 34.

XXV.

1. If the line which bisects the vertical angle of a triangle also bisects the base the triangle is isosceles.

2. Prove that there is one and only one point which is equidistant from three given points not in the same straight line.

3. Show that four equal right-angled isosceles triangles can be arranged round one common vertex so as to form a square.

4. In the triangle ABC the side AB is bisected at M and MN is drawn parallel to BC to meet AC in N. Prove, by drawing NP parallel to AB, that MN bisects the side AC.

5. Any point P is taken in the base BC of an isosceles triangle, and PM and PN are drawn parallel to the equal sides AB, AC to meet them in M and N. Prove that the sum of PM and PN is constant.

6. ABC is any triangle and CPQ is drawn perpendicular to the bisector of the angle A meeting it in P, and the side AB in Q. Prove that each of the angles AQC and ACQ is equal to half the sum of the angles ABC and ACB.

XXVI.

1. If through any point equidistant from two parallel straight lines, two other straight lines be drawn cutting the parallel straight lines, one in the points A and C, the other in B and D, prove that AC=DB.

2. Prove that there are four and only four points in a plane, each of which is equidistant from the three sides of a triangle.

3. Show that six equilateral triangles can be arranged round one common vertex so as to make a regular hexagon.

4. In the triangle ABC the two medians BY and OZ are drawn to intersect in O, and through C, CE is drawn parallel to BY. Join AO, and produce it to meet BC in X, and CE in E. Join BE. Prove that AE is bisected in O, that BOCE is a parallelogram, and that AX is the third median of the triangle ABC.

5. Draw a straight line through a given point, so that the part of it intercepted between two given parallels may be of a given length.

6. AB is a given finite straight line. Draw AC making the angle BAC acute. Produce AC to D, and then to E, making CD and CE each equal to AC. Join BE, and draw CP and DQ parallel to BE. Prove that AB is trisected in the points P and Q.

XXVII.

1. Any line AX is drawn through the angle A of the parallelogram ABCD, and BP, CQ and DR are drawn perpendicular to AX. If C is the angle opposite A,

prove that CQ is equal to the sum or difference of BP and DR, according as AX falls without or intersects the parallelogram.

2. If two straight lines are parallel to two other straight lines, each to each, then the angles contained by the first pair are equal to the angles contained by the other pair.

3. Show that eight equal triangles can be arranged round one common vertex so as to form a regular octagon.

4. Bisect AC and AB the sides of the triangle ABC at the points Y, Z; and draw AP perpendicular to BC. Prove that the angle YPZ is equal to the angle BAC.

5. If two parallelograms have two adjacent sides of the one equal to two adjacent sides of the other, each to each, and one angle of one equal to one angle of the other, the two parallelograms are equal in all respects.

6. If the angle between two adjacent sides of a parallelogram be increased, but the lengths of the sides remain the same, the diagonal through their point of intersection will be diminished.

XXVIII.

1. In a given straight line find a point which is equidistant from two given straight lines. When is this impossible?

2. Half the base of a triangle is greater than, equal to, or less than the line joining the vertex to the middle point of the base, according as the vertical angle is obtuse, right or acute.

3. Find the locus of a point which is at a given distance from a given straight line.

4. In a right-angled triangle ABC, having the right angle ACB, if the angle CAB is double of the angle ABC, then AB is double of AC.

5. Two right-angled parallelograms are equal if two adjacent sides of the one are equal to two adjacent sides of the other, each to each.

6. Bisect AB, CD, two opposite sides of a parallelogram ABCD at M and N. Join CM and NB. Prove that DM and NB trisect the diagonal AC

XXIX.

1. Prove by the method of superposition that if the four angles of a quadrilateral figure are all equal, its opposite sides are equal.

2. If in the sides AB, AC of the triangle ABC, in which AC is greater than AB, points D, E be taken so that BD and CE are equal, prove that CD is greater than BE.

3. Draw a straight line which shall make equal angles with two given intersecting straight lines and be equidistant from two given points.

4. In the figure of Prop. 1 produce AB both ways to meet the circumferences in D and E. Join CD, CE. Prove that CDE is an isosceles triangle having one angle four times each of the other angles.

5. The angle ABC of the triangle ABC is bisected by BD, which meets AC in D; and through D, DE

and DF are drawn parallel to AB and BC. Prove that DEBF is a rhombus.

6. Show how to divide a given straight line into seven equal parts.

XXX.

1. Prove Prop. 27 by the method of superposition.

2. ABC is an isosceles triangle having AB equal to AC. Bisect the angles ABC, ACB by the lines BX and CY meeting AC and AB in X and Y. Prove that the triangles YBC and XCB are equal in all respects.

3. Find the locus of the middle points of all the straight lines drawn from a given point to meet a given straight line of unlimited length.

4. If an exterior angle of a triangle be bisected and also one of the interior opposite angles, the angle contained by the bisecting lines is equal to half the other interior opposite angle of the triangle.

5. If the angular points of one parallelogram lie on the sides of another parallelogram, the diagonals of both parallelograms pass through the same point.

6. Find in a side of a triangle the point from which the straight lines drawn parallel to the other sides of the triangle and terminated by them are equal.

PART VI.

TO EUCLID I. 41.

XXXI.

1. AC and BC are two given straight lines. Show how to draw a straight line from a given point P to AC, so that it is bisected by BC.

2. AB and CD are two straight lines intersecting in O, and X is a given finite straight line. Show how to find the points in AB which are at a perpendicular distance equal to X from CD.

3. The area of any parallelogram is equal to the product of the base into the altitude.

4. ABC is a triangle and D any point in AB. Show how to draw through D a straight line DE to meet BC produced in E, so that the triangle DBE may be equal to the triangle ABC.

5. A triangle is divided by each of its medians into two triangles of equal area.

6. The straight line which joins the middle points of two sides of a triangle is parallel to the third side.

XXXII.

1. The straight lines drawn through the middle points of the sides of a triangle perpendicular to the sides meet in a point.

2. The straight line which joins the middle points of two sides of a triangle is equal to half the third side.

3. Show how to bisect a triangle by a straight line drawn through one of its angular points.

4. If any point O be taken on the median AX of the triangle ABC, prove that the triangle AOB will be equal to the triangle AOC.

5. The area of a triangle is equal to half the product of the base into the altitude.

6. PQRS is a quadrilateral figure. On the base PQ show how to construct a triangle equal in area to PQRS and having the angle at P common with the quadrilateral figure.

XXXIII.

1. Any point on the line which bisects a given rectilineal angle is equidistant from the two lines containing the angle.

2. The three straight lines which join the middle points of the sides of a triangle divide the triangle into four triangles which are equal in all respects.

3. The four triangles into which a parallelogram is divided by its diagonals are equal in area.

4. ABCD is a quadrilateral figure, and X, Y, Z, W are the middle points of its sides. Prove that the figure formed by joining the middle points is a parallelogram whose area is half that of the quadrilateral ABCD.

5. Given the area and the base of a triangle, find the locus of its vertex.

6. A triangle is equal in area to the sum or difference of two triangles on the same base, if the altitude of the former is equal to the sum or difference of the altitudes of the latter.

XXXIV.

1. The bisectors of the angles of a triangle are concurrent.

2. If two sides of a quadrilateral figure are parallel, but unequal, the straight line which joins the middle points of the oblique sides is equal to half the sum of the parallel sides; and the part of this line which is intercepted between the diagonals of the quadrilateral figure is equal to half the difference of the parallel sides.

3. If the diagonals of a quadrilateral figure divide it into four equal triangles it is a parallelogram.

4. ABCD is a parallelogram and E is the middle point of CD. Prove that the triangle AEB will be half the parallelogram.

5. P is any point in BC, the base of the triangle ABC, and X is the middle point of BC. Join AP and draw XQ parallel to AP to meet one of the other sides of the triangle in Q. Join PQ and prove that it bisects the triangle.

6. A triangle is equal in area to the sum or difference of two triangles of the same altitude if the base of the former is equal to the sum or difference of the bases of the latter.

XXXV.

1. ABC is a given triangle. Show how to describe another triangle PQR having the points A, B, C as the middle points of its sides.

2. If ABC and ABD be two equal triangles on the same base AB but on opposite sides of it, prove that they have equal altitudes and that the line joining the vertices C, D is bisected by AB.

3. If two opposite sides of a quadrilateral figure are parallel the straight line which joins the middle points of these two sides will bisect the figure.

4. The three medians of a triangle are concurrent, and cut one another in a point such that one part of any median is double of the other part.

5. The base BC of the triangle ABC is trisected at X and Y, and P is any other point in BC. Show how to draw two lines through P trisecting the triangle ABC.

6. The diagonals of the quadrilateral figure ABCD intersect in O, and OB is produced to E, making OE equal to the diagonal BD. Prove that the triangle AEC is equal in area to the figure ABCD.

XXXVI.

1. Assuming that the three straight lines drawn at right angles to the sides of a triangle at their middle points are concurrent, prove that the three straight lines drawn from the angular points of a triangle perpendicular to the opposite sides are also concurrent.

2. If the line joining the vertices of two triangles on the same base but on opposite sides of it be bisected by the base, the triangles are equal.

3. Bisect a quadrilateral figure by a line drawn through a given vertex.

4. Find a point which shall be at a given distance X from two given intersecting straight lines.

5. When two sides AB, AC of a triangle are given in length the area is a maximum when BAC is a right angle.

6. Two quadrilateral figures are equal when their diagonals are equal and intersect at the same angle.

PART VII.

TO EUCLID I. 48.

XXXVII.

1. A given straight line AB is bisected at C, and perpendiculars AX, CZ, BY are drawn to any other straight line. Show that the projections of AC and CB, that is XZ and ZY, are equal.

2. AB and CD are two straight lines which cut one another, and X is a given finite straight line. Show how to describe an equilateral triangle having its base on AB, its vertex on CD and each of its sides equal to X.

3. Describe a triangle equal to a given parallelogram and having an angle equal to a given rectilineal angle.

4. In the figure of Prop. 47 join FD, EK and GH. Prove that the triangles ABC, FBD, GAH and KCE are all equal.

5. In a rhombus the squares of the four sides are together equal to the squares of the diagonals.

6. Show how to draw through one of the corners of a square two lines which shall divide the square into three equal parts

XXXVIII.

1. Describe a right-angled triangle having its hypothenuse equal to a given finite straight line and the sum of its sides equal to another given finite straight line.

2. The diagonals of parallelograms about a diagonal of a parallelogram are parallel.

3. If three parallel straight lines make equal intercepts on a fourth straight line which meets them, they will also make equal intercepts on any other straight line which meets them.

4. In the figure of Prop. 47 prove that AD and FC are perpendicular.

5. Given the diagonal of a square construct the square.

6. Any point P is taken in the base AB of a triangle, and PQ and PR are drawn parallel to the sides of the triangle, meeting them in Q and R. Prove that the parallelogram PQCR is greatest when P is taken at the middle point of AB.

XXXIX.

1. If two lines be respectively perpendicular to two others, the angle between the former is equal to the angle between the latter.

2. Construct a rectilineal figure equal to a given rectilineal figure, and having fewer sides by one than the given figure.

3. Equal and parallel straight lines have equal projections on any other straight line.

4. ABC is a right-angled triangle having ABC the right angle. On AB, on the side away from C, describe the square ABDE; and on AC, on the same side as B, describe the square ACFG. Draw FH and FK perpendicular to BD and ED produced. Prove that (1) G lies in DE, (2) the triangles ABC, AEG, CHF, GKF are all equal, (3) HK is a square and is equal to the square of BC.

5. On a given base describe a triangle equal to a given triangle.

6. From any point O perpendiculars OX, OY, OZ are drawn to the sides BC, CA, AB of the triangle ABC. Prove that the squares of AZ, BX, CY are together equal to the squares of AY, CX, BZ.

XL.

1. The straight line which is drawn through the middle point of one side of a triangle parallel to another side will bisect the third side of the triangle.

2. Construct a triangle equal to a given rectilineal figure.

3. A given straight line AB is bisected at C, and perpendiculars AX, CZ, BY are drawn to any other straight line PQ, which does not pass between A and B. Prove that CZ is equal to half the sum of AX and BY.

4. A square is described on a line DE which is equal to the sum of the two sides of the right-angled triangle ABC, and from the four corners of this square four right-angled triangles, each identically equal to ABC, are cut away. Prove that the figure left is equal to the square of the hypothenuse AC.

5. Through the point O within the parallelogram ABCD two straight lines are drawn parallel to the sides. If the parallelograms OB and OD are equal prove that O lies on the diagonal of AC.

6. If points X, Y, Z be taken on the sides BC, CA, AB of the triangle ABC, such that the squares of AZ, BX, CY are together equal to the squares of AY, CX, BZ, prove that the perpendiculars to the sides of the triangle at the points X, Y, Z are concurrent.

XLI.

1. If the middle points of the adjacent sides of any quadrilateral figure be joined, prove that the figure thus formed is a parallelogram.

2. OEC is a triangle and the median CX is produced to B so that CX is equal to XB, and EO is produced to A so that EO is equal to OA. Prove that ABC will be a triangle having its medians equal to one and a half times the various sides of the triangle OEC.

3. ABC, ABD are on the same base AB and between the same parallels. A line parallel to AB cuts AC, BC, AD and BD in E, F, G and H. Prove by *reductio ad absurdum* that EF=GH.

4. A square is described on a line DE which is equal to the sum of the two sides of the right-angled triangle ABC, and from two opposite corners of this square two right-angled parallelograms, each double of the triangle ABC, are cut away. Prove that two squares may be left equal respectively to the squares on AB and BC.

5. The square described on the diagonal of a given square is double of the given square.

6. If the opposite angles of a quadrilateral figure are supplementary, a point can be found which is equidistant from the four vertices.

XLII.

1. Of all triangles having the same base and area, the perimeter of an isosceles triangle is least.

2. Show how to construct a triangle having its medians equal to three given lines.

3. If the base BC of a triangle ABC be divided into any number of equal parts at the points P, Q, R, and these points be joined to the vertex A, show that any line parallel to BC will be divided into equal parts by the lines AP, AQ, AR.

4. Show that Prop. 47 may be proved by cutting off four right-angled triangles from each of two equal squares.

5. ABC is an equilateral triangle, and AD is drawn perpendicular to BC. Prove that the square on AD is equal to three times the square on BD or CD.

6. If the sum of one pair of opposite sides of a convex quadrilateral figure is equal to that of the other two sides, a point can be found which is equidistant from the four sides.

PART VIII.

TO EUCLID I. 48.

XLIII.

1. Given four lines, no two of which are parallel. In how many points will these lines intersect, and how many diagonals can be drawn joining two points of intersection. Draw such a complete quadrilateral, and name its sides and diagonals.

2. O is any point outside the parallelogram ABCD, and also outside the angle BAD and its opposite vertical angle. Prove that the triangle DAC will be equal to the sum of the triangles OAD, OAB.

3. Assuming the rider in XLII. 3, show how to divide a given straight line into any given number of equal parts.

4. In a right-angled triangle if a perpendicular be drawn from the right angle to the base, the square on either of the sides containing the right angle is equal to the rectangle contained by the base, and its segment adjacent to that side.

5. AB and CD are two given finite straight lines. Draw BE at right angles to AB, and equal to CD. Show how to find a point H in AB, such that the

difference of the squares on AH and HB shall be equal to the square on CD.

6. In any triangle if a perpendicular be drawn from one extremity of the base to the bisector of the vertical angle, the line joining the middle point of the base to the foot of this perpendicular is equal to half the difference of the sides of the triangle.

XLIV.

1. ABCD is a quadrilateral figure. Two of its opposite sides AD and BC are bisected at X and Y; and its diagonals AC and BD are bisected at Z and W. Prove that XZWY is a parallelogram whose area is equal to half the difference of the areas of the triangles ABC and ABD.

2. ABCD is a parallelogram, and O is any point within the angle BAD or its opposite vertical angle. Prove that the triangle OAC is equal to the difference of the triangles OAD, OAB.

3. Any straight line drawn from the vertex of a triangle to the base is bisected by the straight line which joins the middle points of the other sides of the triangle.

4. ABC is a right-angled isosceles triangle, having the side AB equal to BC. If BC is produced to D, E and F, making BD equal to AC, BE equal to AD, and BF equal to AE, show that the squares on BD, BE and BF are equal to twice, three times and four times the square on AB.

5. Given the base of a triangle, and the difference of the squares on the sides of the triangle. Show that

the locus of the vertex of the triangle is a straight line perpendicular to the base.

6. Assuming the last rider, No. 5, prove that the three perpendiculars from the angles of a triangle to the opposite sides are concurrent.

XLV.

1. Assuming the figure and rider XLIV. 1, if AB and DC meet in L, and ZY produced meets LC in K; prove that each of the triangles LZY and CZY is equal to one-fourth of ABC; that each of the triangles LWY and BWY is equal to one-fourth of BCD; and that the triangle LZW is equal to one-fourth of the quadrilateral figure ABCD.

2. In a triangle ABC, AD is drawn perpendicular to BC, and X, Y, Z are the middle points of the sides BC, CA, AB. Prove that each of the angles ZXY, ZDY is equal to the angle BAC.

3. If two straight lines AB, CD intersect in O, so that the triangle AOC is equal to the triangle DOB, prove that AD and CB are parallel.

4. Show how to divide a given straight line into two parts, so that the square of one part may be double of the square of the other part.

5. In the figure of Prop. 47 join FD and EK, and prove that the square on FD is equal to the square on AB together with four times the square on AC.

6. Construct a square so that one side shall lie on a given straight line and two other sides shall pass through two given points.

XLVI.

1. Assuming XLV. 1, show that the middle points of the three diagonals of a complete quadrilateral are collinear, *i.e.* in the same straight line.

2. The perpendiculars through the middle points of the sides of a triangle ABC meet in P, and the medians meet in M. Join PM and produce it to meet AD, the perpendicular from A to BC, in O. Prove that MO = twice PM, and that all the perpendiculars from the angles of the triangle pass through O.

3. The angles ABC and ACB are bisected by the lines BK and CK, and DKE is drawn through K parallel to BC meeting AB and AC in D and E. Prove that DE is equal to the sum of BD and CE.

4. Show how to divide a given straight line into two parts so that the square of one part may be equal to three times the square of the other part.

5. In the figure of Prop. 47 join FD and EK and prove that the squares on FD and EK are equal to five times the square on BC.

6. Construct a square so that two opposite sides shall pass through two given points, and its diagonals intersect at a third given point.

XLVII.

1. The quadrilateral figure, which is formed by the four straight lines bisecting the angles of any quadrilateral figure ABCD, has its opposite angles equal to two right angles.

2. The perpendiculars through the middle points of the sides of a triangle ABC meet in P, and the perpendiculars to the sides from the angles of the triangle meet in O. Prove that AO is equal to twice the length of the perpendicular from P on BC.

3. Having given the direction of two lines AB and AC, and that BC always passes through a given point P, prove that the triangle ABC will be least when BC is bisected at P.

4. Having given one side of a right-angled parallelogram which is equal to a given square, find the length of the other side of the parallelogram.

5. The triangle formed by the three bisectors of the exterior angles of a triangle is such that the lines joining its vertices to the angles of the original triangle will be its perpendiculars.

6. The equilateral triangle described on the hypothenuse of a right-angled triangle is equal to the sum of the equilateral triangles described on the sides.

XLVIII.

1. The quadrilateral figure, which is formed by the four straight lines bisecting the exterior angles of any quadrilateral figure ABCD, has its opposite angles equal to two right angles.

2. The point of concurrence of the perpendiculars to the sides of a triangle at their middle points, the point of concurrence of the perpendiculars to the sides from the opposite angles, and the point of concurrence of the medians are collinear.

3. Having given two lines AB and CD not parallel to each other, find the straight line which would bisect the angle between AB and CD without producing them.

4. In every quadrilateral the intersection of the straight lines which join the middle points of opposite sides is the middle point of the straight line which joins the middle points of the diagonals.

5. If the opposite angles of a quadrilateral figure are equal, the opposite sides are equal.

6. On AB, BC the sides of a triangle ABC, any parallelograms ABFE, BCDL are constructed, and EF, DL are produced to meet in O. On AC a parallelogram ACHG is constructed having AG, CH equal and parallel to OB. Prove that it is equal to the sum of the other two parallelograms.

PART IX.

ON EUCLID, BOOK II.

XLIX.

1. State the first three Propositions in Bk. II. in Algebraical Formulae, and show that Props. 2 and 3 are only special cases of Prop. 1.

2. Show by Prop. 4 that the square on any straight line is equal to four times the square on half the line.

3. Write out a full geometrical proof that $a^2 - b^2 = (a+b)(a-b)$.

4. In Prop. 11 produce EC to G making EG equal to BE, and on AG describe a square having one corner in AB produced in K. Prove that the rectangle AB, AK is equal to the square on BK.

5. The difference of the squares on two sides of a triangle is equal to twice the rectangle contained by the base and that part of the base intercepted between the middle point of the base and the foot of the perpendicular drawn from the opposite angle.

6. The sides of a triangle are 10, 12, 15 inches. Prove that it is acute-angled.

L.

1. State Props. 4, 5, 6, 7 in algebraical formulae.

2. Prove Prop. 4 by means of Props. 2 and 3.

3. If a line is divided into any two parts the rectangle contained by the parts is a maximum, and the sum of their squares is a minimum, when the parts are equal.

4. The square on a straight line AD drawn from the vertex A of an isosceles triangle to any point D in the base is less than the square on AB, one of the equal sides, by the rectangle contained by BD, DC the segments of the base.

5. If a straight line be divided internally in medial section as in Prop. 11, and if from AH the greater segment, a part be taken equal to HB the less, show that AH the greater segment is also divided in medial section.

6. The sum of the squares on the sides of a parallelogram is equal to the sum of the squares on the diagonals.

LI.

1. If any four points A, B, C, D are taken in order along a straight line, prove that
$$AB,CD + BC,AD = AC,BD$$ and that this is the same as $AB,CD + BC,AD + CA,BD = 0$.

2. In a right-angled triangle, if a perpendicular is drawn from the right angle to the hypothenuse, the square on this perpendicular is equal to the rectangle contained by the segments of the hypothenuse.

3. A and B are two fixed points, and the point D moves so that the difference of the squares on AD and

DB is constant. Prove that the locus of D is a straight line perpendicular to the line passing through A and B.

4. State Props. 9, 10, 12, 13 in algebraical formulae.

5. Deduce Props. 9 and 10 from Props. 4 and 7.

6. In Prop. 11 show that the rectangle contained by the sum and difference of the parts is equal to the rectangle contained by the parts.

LII.

1. State and prove geometrically that
$$a(b-c)+b(c-a)+c(a-b)=0.$$

2. In a right-angled triangle if a perpendicular be drawn from the right angle to the hypothenuse, the square on either of the two sides containing the right angle is equal to the rectangle contained by the hypothenuse and the segment of it adjacent to that side.

3. The square on the difference of two lines is less than the sum of the squares on those lines by twice the rectangle contained by them.

4. The sum of the squares of the distances of a point D from two given points A and B is constant. Prove that the locus of D is a circle whose centre is the midpoint of the line joining A and B.

5. State Prop. 11 as a quadratic equation. Solve it and explain the two solutions.

6. In any triangle the sum of the squares on two sides is equal to twice the square on half the third side together with twice the square on the median which bisects the third side.

LIII.

1. State and prove geometrically that $(a+b+c)^2 = a^2+b^2+c^2+2ab+2ac+2bc$.

2. Of all rectangles of the same perimeter the square has the greatest area.

3. Prove Prop. 8 by means of Props. 4 and 7.

4. Show how to divide a given straight line into two parts so that the difference of the squares on the parts may be equal to a given square.

5. If a line AB be divided in C so that the rectangle AB, BC is equal to the square on AC, prove that the sum of the squares on AB and BC is equal to three times the square on AC.

6. Three times the sum of the squares of the sides of a triangle is equal to four times the sum of the squares of its three medians.

LIV.

1. In any quadrilateral figure the squares on the diagonals are together equal to twice the sum of the squares on the straight lines joining the middle points of adjacent sides.

2. If a line AB is divided equally at C and unequally at D, prove that the difference of the squares on AD, DB is equal to twice the rectangle AB, CD.

3. The line AB is divided into any two parts at C, and produced to D, making BD equal to BC. Show that four rectangles, each equal to the rectangle AB, BC can be cut from the square on AD, one rectangle

being taken at each of its corners, so as to leave a square equal to the square of AC.

4. Show how to produce a given straight line so that the rectangle contained by the whole line thus produced and the part produced may be equal to the square of the original line.

5. If a line AB be divided at C so that the rectangle AB, BC is equal to the square of AC, prove that $(AC+BC)^2 = 5AC^2$.

6. The sum of the squares of the four sides of a quadrilateral figure is equal to the sum of the squares of its diagonals plus four times the square of the line joining the middle points of the diagonals.

PROPOSITIONS IN EUCLID

Which may be Set to be written out with the Riders.

NUMBER OF RIDER.		1	2	3	4	5	6
Part I.	Paper I.	...	...	1	8	2	...
,,	,, II.	...	...	1	4	2	9
,,	,, III.	...	...	8	4	4	..
,,	,, IV.	...	...	8	6	4	...
,,	,, V.	...	5	8	8	4	9
,,	,, VI.	12	5	4	8	5	...
Part II.	Paper VII.	1	8	4	13	16	20
,,	,, VIII.	4	8	4	14	19	20
,,	,, IX.	11	8	9	15	18	20
,,	,, X.	11	4	15	20	21	4
,,	,, XI.	11	4	20	8	3	4
,,	,, XII.	11	4	23	26	3	4
Part III.	Paper XIII.	5	26	27	28	29	32
,,	,, XIV.	4	26	5	29	32	32
,,	,, XV.	2	4	29	23	31	32
,,	,, XVI.	2	26	4	29	23	32
,,	,, XVII.	17	21	26	28	23	32
,,	,, XVIII.	16	8	26	28	32	1
Part IV.	Paper XIX.	9	22	32	23	32	34
,,	,, XX.	33	22	32	6	32	4
,,	,, XXI.	18	32	32	24	32	34
,,	,, XXII.	18	22	26	29	8	29
,,	,, XXIII.	32	26	32	26	28	29
,,	,, XXIV.	22	26	8	4	34	34

RIDER PAPERS.

NUMBER OF RIDER.		1	2	3	4	5	6
Part v.	Paper xxv.	6	4	15	26	34	32
,,	xxvi.	26	4	15	34	31	29
,,	xxvii.	26	29	15	32	34	24
,,	xxviii.	26	24	34	32	4	29
,,	xxix.	8	24	9	1	6	29
,,	xxx.	8	26	31	32	34	9
Part vi.	Paper xxxi.	31	36	35	37	38	39
,,	xxxii.	4	34	38	38	41	37
,,	xxxiii.	26	34	38	30	37	41
,,	xxxiv.	26	34	38	41	37	41
,,	xxxv.	31	34	38	29	37	38
,,	xxxvi.	34	38	39	39	41	37
Part vii.	Paper xxxvii.	34	40	42	4	47	38
,,	xxxviii.	32	29	34	47	46	43
,,	xxxix.	32	37	34	47	44	47
,,	xl.	29	37	34	47	45	48
,,	xli.	30	37	37	47	47	18
,,	xlii.	37	22	38	47	47	26
Part viii.	Paper xliii.	...	41	38	47	48	26
,,	xliv.	38	41	38	47	48	48
,,	xlv.	38	32	39	47	47	46
,,	xlvi.	38	26	29	47	47	46
,,	xlvii.	32	26	4	43	34	47
,,	xlviii.	32	26	9	34	28	47
Book II.							
Part ix.	Paper xlix.	1	4	5	11	12	13
,,	l.	7	3	5	12	11	13
,,	li.	1	4	5	9	7	11
,,	lii.	1	3	5	10	11	13
,,	liii.	4	5	7	5	13	11
,,	liv.	4	5	6	11	11	13

ENUNCIATIONS OF THE PROPOSITIONS IN EUCLID.

BOOK I.

1. To describe an equilateral triangle on a given finite straight line.

2. From a given point to draw a straight line equal to a given straight line.

3. From the greater of two given straight lines to cut off a part equal to the less.

4. If two triangles have two sides of the one equal to two sides of the other, each to each, and have also the angles contained by those sides equal to one another, they shall also have their bases or third sides equal; and the two triangles shall be equal, and their other angles shall be equal, each to each, namely those to which the equal sides are opposite.

5. The angles at the base of an isosceles triangle are equal to one another, and, if the equal sides be produced, the angles on the other side of the base shall be equal to one another.

6. If two angles of a triangle be equal, the sides also which subtend, or are opposite to the equal angles, shall be equal to one another.

7. On the same base and on the same side of it there cannot be two triangles having their sides, which are terminated in one extremity of the base, equal to one

another, and likewise those which are terminated at the other extremity, equal to one another.

8. If two triangles have two sides of the one equal to two sides of the other, each to each, and have likewise their bases equal, the angle which is contained by the two sides of the one shall be equal to the angle which is contained by the two sides, equal to them, of the other.

9. To bisect a given rectilineal angle—that is, to divide it into two equal parts.

10. To bisect a given finite straight line—that is, to divide it into two equal parts.

11. To draw a straight line at right angles to a given straight line from a given point in the same.

12. To draw a straight line perpendicular to a given straight line of unlimited length from a given point without it.

13. The angles which one straight line makes with another straight line on one side of it are either two right angles or are together equal to two right angles.

14. If at a point in a straight line two other straight lines on opposite sides of it make the adjacent angles together equal to two right angles, these two straight lines shall be in one and the same straight line.

15. If two straight lines cut one another, the vertical or opposite angles are equal. .

16. If one side of a triangle be produced, the exterior angle shall be greater than either of the interior and opposite angles.

17. Any two angles of a triangle are together less than two right angles.

18. The greater side of every triangle has the greater angle opposite to it.

19. The greater angle of every triangle is subtended by the greater side or has the greater side opposite it.

20. Any two sides of a triangle are together greater than the third side.

21. If from the ends of the side of a triangle there be drawn two straight lines to a point within the triangle, these shall be less than the other two sides of the triangle, but shall contain a greater angle.

22. To make a triangle of which the sides shall be equal to three given straight lines, any two of which are together greater than the third.

23. At a given point in a given straight line to make an angle equal to a given rectilineal angle.

24. If two triangles have two sides of the one equal to two sides of the other, each to each, but the angle contained by the two sides of one of them greater than the angle contained by the two sides equal to them of the other, the base of that which has the greater angle shall be greater than the base of the other.

25. If two triangles have two sides of the one equal to two sides of the other, each to each, but the base of the one greater than the base of the other, the angle contained by the sides of that which has the greater base, shall be greater than the angle contained by the sides, equal to them, of the other.

26. If two triangles have two angles of the one equal to two angles of the other, each to each, and one side equal to one side, namely, either the sides adjacent to the equal angles or sides which are opposite to equal angles in each, then shall the other sides be equal, each to each, and also the third angle of the one equal to the third angle of the other.

27. If a straight line falling on two other straight lines makes the alternate angles equal to one another, the two straight lines shall be parallel.

28. If a straight line falling on two other straight lines makes the exterior angle equal to the interior and opposite angle on the same side of the line, or makes the interior angles on the same side together equal to two right angles, the two straight lines shall be parallel.

29. If a straight line fall on two parallel straight lines, it shall make the alternate angles equal, and the exterior angle equal to the interior and opposite angle on the same side; and also the two interior angles on the same side together equal to two right angles.

30. Straight lines which are parallel to the same straight line are parallel to one another.

31. To draw a straight line through a given point parallel to a given straight line.

32. If a side of any triangle be produced, the exterior angle is equal to the two interior and opposite angles, and the three interior angles of every triangle are together equal to two right angles.

33. The straight lines which join the extremities of two equal and parallel straight lines towards the same parts are themselves equal and parallel.

34. The opposite sides and angles of a parallelogram are equal to one another, and the diameter bisects the parallelogram, *i.e.* divides it into two equal parts.

35. Parallelograms on the same base and between the same parallels are equal to one another.

36. Parallelograms on equal bases and between the same parallels are equal to one another.

37. Triangles on the same base and between the same parallels are equal to one another.

38. Triangles on equal bases and between the same parallels are equal to one another.

39. Equal triangles on the same base and on the same side of it are between the same parallels.

40. Equal triangles on equal bases in the same straight line and on the same side of it are between the same parallels.

41. If a parallelogram and a triangle be on the same base and between the same parallels, the parallelogram shall be double of the triangle.

42. To describe a parallelogram that shall be equal to a given triangle, and have one of its angles equal to a given rectilineal angle.

43. The complements of the parallelograms which are about the diameter of any parallelogram are equal to one another.

44. To a given straight line to apply a parallelogram which shall be equal to a given triangle, and have one of its angles equal to a given rectilineal angle.

45. To describe a parallelogram equal to a given rectilineal figure, and having an angle equal to a given rectilineal angle.

46. To describe a square on a given straight line.

47. In any right-angled triangle, the square which is described on the side subtending the right angle is equal to the squares described on the sides containing the right angle.

48. If the square described on one of the sides of a triangle be equal to the squares described on the other two sides, the angle contained by these two sides is a right angle.

BOOK II.

1. If there be two straight lines, one of which is divided into any number of parts, the rectangle contained by the two straight lines is equal to the rectangles contained by the undivided line and the several parts of the divided line.

2. If a straight line be divided into any two parts, the rectangles contained by the whole and each of the parts are together equal to the square on the whole line.

3. If a straight line be divided into any two parts, the rectangle contained by the whole and one of the parts is equal to the rectangle contained by the two parts, together with the square on the aforesaid part.

4. If a straight line be divided into any two parts, the square on the whole line is equal to the squares on the two parts, together with twice the rectangle contained by the two parts.

ENUNCIATIONS OF PROPOSITIONS.

5. If a straight line be divided into two equal parts, and also into two unequal parts, the rectangle contained by the unequal parts, together with the square on the line between the points of section, is equal to the square on half the line.

6. If a straight line be bisected and produced to any point, the rectangle contained by the whole line thus produced and the part produced, together with the square on half the line bisected, is equal to the square on the straight line which is made up of the half and the part produced.

7. If a straight line be divided into any two parts, the squares on the whole line and on one of the parts are equal to twice the rectangle contained by the whole and that part, together with the square on the other part.

8. If a straight line be divided into any two parts, four times the rectangle contained by the whole line and one of the parts, together with the square on the other part, is equal to the square on the straight line which is made up of the whole and that part.

9. If a straight line be divided into two equal and into two unequal parts, the squares on the two unequal parts are together double of the square on half the line, and of the square on the line between the points of section.

10. If a straight line be bisected and produced to any point the squares on the whole line thus produced and the part produced are together double of the square on half the line and of the square on the line made up of the half and the part produced.

11. To divide a given straight line into two parts, so that the rectangle contained by the whole and one part shall be equal to the square on the other.

12. In obtuse-angled triangles, if a perpendicular be drawn from either of the acute angles to the opposite side produced, the square on the side subtending the obtuse angle is greater than the squares on the sides containing the obtuse angle by twice the rectangle contained by the side on which, when produced, the perpendicular falls and the straight line intercepted without the triangle between the perpendicular and the obtuse angle.

13. In every triangle the square on the side subtending an acute angle is less than the squares on the sides containing that angle by twice the rectangle contained by either of these sides and the straight line intercepted between the perpendicular let fall on it from the opposite angle and the acute angle.

14. To describe a square that shall be equal to a given rectilineal figure.

EXAMINATION PAPERS IN EUCLID.

I.
College of Preceptors, Midsummer 1890.
Third Class.

1. What meaning do you give to the terms *base, radius, parallelogram?*

    Write out one Postulate and two Axioms.

2. Euclid I. 2.

    Suppose A were on the circumference of the smaller circle used in the construction, where would the vertex of the equilateral triangle fall?

3. Euclid I. 5.

    Show that the straight line which bisects the vertical angle of an isosceles triangle also bisects the base.

4. Euclid I. 12.

5. Any two angles of an isosceles triangle are together less than two right angles.

6. Euclid I. 19.

7. *Either*, Euclid I. 25.

    Or, If two straight lines cut one another, and if two of the adjacent angles be bisected, the bisecting lines shall be at right angles to one another.

II.
College of Preceptors, Christmas 1890.
Third Class.

1. Define a *straight line*, a *plane rectilineal angle*, and an *equilateral triangle*.

2. Euclid I. 1.
3. Euclid I. 6.
4. Euclid I. 9.

If D and E are points on AB and AC equidistant from A, show that the bisector of the angle bisects DE and is at right angles to it.

5. Euclid I. 13.

Two straight lines AC, AD are drawn from A in the line BAE, on one side of it; then the angles BAC, CAD, DAE together equal two right angles.

6. Euclid I. 18.
7. *Either*, Euclid I. 22.

Show how the construction would fail if two of the lines were not together greater than the third. [Illustrate your answer by a figure.]

Or, Euclid I. 26, Case 1.

III.

College of Preceptors, Midsummer 1890.
Second Class.

1. Define a *point* and a *straight line*.

Euclid I. 2. How would you proceed further to draw a second equal straight line from the given point in a given direction?

If the given point lies on the smaller of the two circles (in Euclid's construction), show that the vertex of the equilateral triangle employed lies also on this circle.

2. Define a *triangle*, and classify triangles according to the equality or inequality of their sides.

Prove in any way Euclid I. 8.

Hence show that, if the opposite sides of a quadrilateral are equal, the opposite angles are equal.

3. When is a straight line said to be *at right angles* to a given straight line?

Euclid I. 12.

4. What is meant by "the *exterior* angle of a triangle formed by producing a side of the triangle"? How many such angles are there in a triangle?

Euclid I. 16. Is an exterior angle of a triangle greater or less than the *adjacent interior angle?*

5. Can we form a triangle with *any* three given lengths? Enunciate the Proposition on which you ground your answer.

The sum of the sides of a convex four-sided figure is greater than the sum of its two diagonals. Prove this.

6. Euclid I. 26, Case 1.

7. Euclid I. 38. What is meant by *equal* in the enunciation?

ABCD is a square; BC, CD are bisected in E, F, respectively; and AE, EF, AF are drawn. Prove that △AEF is three-eighths of the square.

8. Euclid I. 42.

9. ABC is a right-angled triangle, A the right angle; squares BDEC, ABFG are described *externally* on BC, BA respectively; and AL is drawn perpendicular to DE to meet it in L. Prove that the rectangle BL equals the square AF. Prove also that AD is perpendicular to FC.

10. (i.) ABC is an equilateral triangle; on BC is described the square BDEC, and on DE the equilateral triangle DEF. Prove that EF is parallel to AB.

Or, (ii.) Bisect a parallelogram by a straight line

drawn through a given point in the plane of the parallelogram.

IV.
College of Preceptors, Christmas 1890.
Second Class.

1. Define *line, obtuse angle, rhombus.*

Name as many different kinds of triangles as you can, with a picture of each.

2. Show how, with a *plain* ruler and a pair of compasses, you can produce a straight line, so as to be three times its original length.

3. PQ is a straight line, and R a point. From R draw a straight line equal to PQ.

Write out the Postulates and Axioms used in the construction.

4. *Either,* Euclid I. 8.

How many parts, at least, of one triangle must be equal to the corresponding parts of another triangle, so that the triangles may be equal in every respect? Draw figures to illustrate your answer.

Or, Euclid I. 13.

If one of the four angles which two intersecting straight lines make with one another, be a right angle, all the others are right angles.

5. Euclid I. 19.

Prove that the hypothenuse of a right-angled triangle is greater than either of the other sides.

6. Define *parallel straight lines.* Write down any Axiom you have learned bearing on the doctrine of parallels.

Prove Euclid I. 27 (after the method of superposition by preference).

Two straight lines perpendicular to the same straight line are parallel.

7. Euclid I. 39.

The sides AB, AC of a triangle ABC are bisected at the points E and F. Prove that EF is parallel to BC. Thence show that if a perpendicular is drawn from A to the opposite side meeting it at D, the angle FDE is equal to the angle BAC. Also show that the figure AFDE is equal to half the triangle ABC.

8. On the base of an equilateral triangle, construct an oblong (or rectangle) equal in area to the triangle.

V.

College of Preceptors, Midsummer 1890.
First Class.

1. Define *line, superficies, polygon, proposition, hypothesis.*

Euclid I. 5.

Prove that a triangle is isosceles, if the bisector of any angle is perpendicular to the opposite side.

2. Euclid I. 14.

3. Euclid I. 21, Part I.

The four sides of any quadrilateral figure are together greater than the two diagonals together.

4. Show that any angle of a triangle is *obtuse, right,* or *acute,* according as it is greater than, equal to, or less than the other two angles of the triangle taken together. Construct an isosceles triangle which shall have the vertical angle four times each of the angles at the base.

5. Euclid II. 6.

(Questions 6, 7, 8, 9 and 10 were set on Books III. and IV.)

VI.
College of Preceptors, Christmas 1890.
First Class.

1. Euclid I. 6.
2. Euclid I. 20.

ABCD is a square: for what position of a point X is the sum of the straight lines XA, XB, XC and XD the least possible? Prove your answer.

3. Euclid I. 32.

The angle contained by one side of a regular polygon and an adjacent side produced is equal to half an angle at the base of an isosceles right-angled triangle. How many sides has the polygon? Explain how you get your result.

4. Euclid I. 48.
5. Euclid II. 5.

Enunciate this Proposition as one about (i.) the rectangle contained by two lines, (ii.) the rectangle contained by the sum and difference of two lines.

6. Divide a straight line AB at the point C, so that the rectangle contained by AB, BC may be equal to the square on AC.

Produce AB to D, making BD equal to BC; then the square on AD is equal to five times the square on AC.

(Questions 7, 8, and 9 were set on Books III. and IV.)

VII.
Oxford Local Examinations, July 1889.
Junior Candidates.

1. Define a *circle*, an *obtuse-angled triangle*, *parallel straight lines*.

2. Euclid I. 10.

In the figure of Euclid I., Prop. 1, if the two points in which the circles meet be joined, the given finite straight line will be bisected.

3. Euclid I. 36.

4. Show that if a quadrilateral be bisected by both its diagonals it is a parallelogram.

5. Euclid I. 48.

6. Euclid II. 6.

7. Prove that if ABC be a triangle, obtuse-angled at B, and D be the foot of the perpendicular from C on AB produced, the square on AC exceeds the squares on AB, BC, by twice the rectangle AB, BD.

(Questions 8, 9, 10, 11 and 12 were set on Books III., IV. and VI.)

VIII.

Cambridge Local Examinations, December 1890.
Junior Students.
Elementary Euclid. Books I., II.

1. Define a *plane superficies*, a *plane rectilineal angle*, and a *right-angled triangle*.

Give Euclid's Axiom relating to right angles.

2. Euclid I. 5.

If on a common base and on opposite sides of it be described two isosceles triangles, the straight line joining their vertices will cut the base at right angles.

3. Euclid I. 34.

The diagonals AC, BD of a quadrilateral ABCD intersect in O, and the parallelograms OAEB, OBFC, OCGD, ODHA are completed: prove that EFGH will be a parallelogram, and will be double of the quadrilateral ABCD.

4. Euclid I. 37.

Through A, B, C are drawn three parallel straight lines to meet the opposite sides of the triangle ABC (produced if necessary) in A', B', C': prove that the triangle A'B'C' will be double the triangle ABC.

5. Euclid I. 48.

6. Euclid II. 5.

7. Euclid II. 11.

Produce a given straight line to a point, such that the rectangle contained by the whole line thus produced and the part produced may be equal to the square on the given straight line.

IX.

Cambridge Local Examinations, December 1890.
Senior Students.

1. Euclid I. 32.

ABC is any acute angle, AB is bisected in D, and at K in BC the angle DKB is made equal to the angle DBK; if AK be drawn, prove that it is perpendicular to BC.

2. Euclid I. 34.

ABCD is a parallelogram, BOD one of its diagonals, and EOG, FOH are drawn parallel to BC, CD respectively, so that E, F, G, H lie, correspondingly, on the sides AB, BC, CD, DA. If DF, BH be drawn intersecting EG in K, L respectively, prove that OK is equal to OL.

3. Euclid II. 14.

A straight line AB is produced both ways to C and D, so that BD is twice AC: show how to find the points C and D when the rectangle CA, AD is equal to the square on AB.

(Questions 4, 5, 6 and 7 were set on Books III., IV., VI. and XI.)

X.
Oxford and Cambridge School Examinations, 1890.
For Commercial Certificates.

1. If two triangles have three sides of the one equal to three sides of the other each to each, the triangles are equal to one another in every respect.

Prove that the diagonals of a rhombus bisect one another, and cut one another at right angles.

2. Euclid I. 22.

Show how the construction would fail if two of the straight lines were together not greater than the third.

3. Euclid I. 32.

Show that each angle of a regular polygon with fifteen sides is twenty-six fifteenths of a right angle.

4. Euclid I. 46.
5. Euclid II. 11.
7. Euclid II. 13.

(Questions 6 and 8 were set on Book III.)

XI.
Oxford and Cambridge School Examinations, 1890.
For Lower Certificates.

1. Define a *parallelogram*, a *plane*, a *circle*.
Euclid I. 7.

ACB, ADB are two triangles on the same side of AB, such that AC is equal to BD and AD is equal to BC, and AD and BC intersect in R ; prove that the triangles ARC and BRD are equal in all respects.

2. Euclid I. 16.

3. Euclid I. 33.

If two sides of a quadrilateral be parallel and unequal and the other two sides be equal, the diagonals are equal.

4. Euclid I. 43.

5. Euclid II. 4.

6. Euclid II. 14.

Divide a given line into two parts so that the rectangle contained by the parts shall be equal to a given square. When is this impossible?

(Questions 7, 8, 9 and 10 were set on Books III., IV. and VI.)

XII.

Oxford and Cambridge School Examinations, 1890.
For Higher Certificates.

1. Euclid I. 10.

Prove that the two straight lines which join the middle points of the sides of an isosceles triangle to the middle point of the base are equal to one another.

2. Euclid I. 29.

The side BC of a triangle ABC is produced to D. Show that the straight lines which bisect the angles BAC, ACD cannot be parallel.

3. Euclid I. 47.

Prove that if the diagonals of a quadrilateral are at right angles the squares on two opposite sides are together equal to the squares on the other two sides.

4. Euclid II. 11.

Prove that if a straight line be divided as above the rectangle contained by the two parts is equal to the difference of the squares on the two parts.

5. Define a *plane superficies*, a *circle*, a *rectilineal figure*.

Show that the distance between the centres of two circles whose circumferences cut one another is less than the sum, and greater than the difference of their radii.

Prove that a quadrilateral cannot have all its angles obtuse.

6. Euclid I. 24.
7. Euclid I. 36.
8. Euclid II. 5.

XIII.

Admission to the R. M. Academy, Woolwich, June 1890.

1. Euclid I. 12.
2. Euclid I. 32.

Draw a straight line DE parallel to the base BC of a triangle to cut the side AB in D and AC in E, so that DE may be equal to BD and CE together.

3. D is a point in the side AB of a triangle. Find a point E in the side BC such that the triangles EAD, CAE may be equal.

4. Euclid I. 47.

Make a square which is three times the square on a given straight line.

5. Euclid II. 11.
6. Give a geometrical proof of the algebraic formula:—
$$(a+b)^2 + (a-b)^2 = 2(a^2+b^2).$$

(Questions 7, 8, 9, 10, 11 and 12 were set on Books III., IV. and VI.)

XIV.

Admission to the R. M. Academy, Woolwich, November 1890.

1. Euclid I. 5.
2. Euclid I. 27.

3. Define a *rhombus;* and show that a rhombus is a parallelogram, and that its diagonals are at right angles.

4. Euclid II. 4.

Show that the area of any rectangle AHGC is half the area of the rectangle contained by the diagonals of the squares described on two adjacent sides of AHGC.

5. Euclid II. 12.

If squares ABDE, ACFG be described outwards on the sides AB, AC of a triangle ABC; show that the sum of the squares on EG and BC is double of the sum of the squares on AB and AC.

(Questions 6, 7, 8, 9, 10, 11 and 12 were set on Books III., IV. and VI.)

XV.

London University Matriculation Examination,
June 1890.

1. Euclid I. 16.
2. Euclid I. 35.
3. Euclid II. 9.

4. If O be any point on the base AC of the isosceles triangle ABC, prove that the rectangle contained by AO and OC is equal to the difference of the squares on AB and OB.

5. If CD be any chord of a circle, P any point on a diameter parallel to CD, and Q the point on the circle which is farthest away from the chord CD, prove that the square on PC and the square on PD are together double the square on PQ.

(Questions 6, 7, 8, 9 and 10 were set on Books III. and IV.)

XVI.

London University Matriculation Examination, January 1891.

1. Prove that the diagonals of a parallelogram bisect each other.

2. Squares are described on the three sides of a right-angled triangle; divide the square on the hypothenuse into two rectangles which shall be respectively equal to the squares on the other sides. (Give the proof.)

3. Euclid I. 22. When is the construction impossible?

Show that if the square on one of the lines exceeds the sum of the squares on the other two, the triangle will have an obtuse angle.

4. Construct a square which shall be equal to a given triangle.

5. Prove that the sum of the squares on the sides of a parallelogram is equal to the sum of the squares on its diagonals.

(Questions 6, 7, 8, 9 and 10 were set on Books III. and IV.)

Messrs. Macmillan and Co.'s Mathematical Books.

BY H. S. HALL, M.A., AND S. R. KNIGHT, M.A.

ELEMENTARY ALGEBRA FOR SCHOOLS. SIXTH EDITION. Globe 8vo. 3s. 6d.; with Answers, 4s. 6d.

ATHENÆUM :—" A work of exceptional merit."

NATURE :—"This is, in our opinion, the best Elementary Algebra for school use. It is the combined work of two teachers who have had considerable experience of actual school teaching . . . and so successfully grapples with difficulties which our present text-books in use, from their authors lacking such experience, ignore or slightly touch upon, . . . We confidently recommend it to mathematical teachers, who, we feel sure, will find it the best book of its kind for teaching purposes."

ACADEMY :—"We will not say that this is the best Elementary Algebra for school use that we have come across, but we can say that we do not remember to have seen a better. . . . It is the outcome of a long experience of school teaching, and so is a thoroughly practical book. . . . Buy or borrow the book for yourselves and judge, or write a better."

KEY. Crown 8vo. 8s. 6d.

HIGHER ALGEBRA. A Sequel to Elementary Algebra for Schools. THIRD EDITION, revised and enlarged. Crown 8vo. 7s. 6d.

** *The third edition contains a collection of three hundred Miscellaneous Examples which will be found useful for advanced Students. These examples have been selected mainly from recent Scholarship or Senate House papers.*

ATHENÆUM :—"We unhesitatingly assert that it is by far the best work of its kind with which we are acquainted. It supplies a want much felt by teachers."

ACADEMY :—". . . Is as admirably adapted for College students as its predecessor was for schools. It is a well-arranged and well-reasoned-out treatise, and contains much that we have not met with before in similar works. . . . The book is almost indispensable and will be found to improve upon acquaintance."

SATURDAY REVIEW :—"They have presented such difficult parts of the subject as Convergency and Divergency of Series, Series generally, and Probability with great clearness and fulness of detail. . . . No student preparing for the University should omit to get this work in addition to any other he may have, for he need not fear to find here a mere repetition of the old story."

KEY. Crown 8vo. 10s. 6d.

MACMILLAN AND CO., LONDON.

Messrs. Macmillan and Co.'s Mathematical Books.

BY H. S. HALL, M.A., AND S. R. KNIGHT, M.A.

ALGEBRAICAL EXERCISES AND EXAMINATION PAPERS. To accompany "Elementary Algebra." Third Edition, revised and enlarged. Globe 8vo. 2s. 6d.

SATURDAY REVIEW:—"To the exercises, one hundred and twenty in number, are added a large selection of examination papers set at the principal examinations, which require a knowledge of algebra. These papers are intended chiefly as an aid to teachers, who no doubt will find them useful as a criterion of the amount of proficiency to which they must work up their pupils before they can send them in to the several examinations with any certainty of success."

IRISH TEACHERS' JOURNAL:—"We know of no better work to place in the hands of junior teachers, monitors, and senior pupils. Any person who works carefully and steadily through this book could not possibly fail in an examination of Elementary Algebra. . . . We congratulate the authors on the skill displayed in the selections of examples."

ARITHMETICAL EXERCISES AND EXAMINATION PAPERS. With an Appendix containing Questions in LOGARITHMS AND MENSURATION. Second Edition. Globe 8vo. 2s. 6d.

CAMBRIDGE REVIEW:—"All the mathematical work these gentlemen have given to the public is of genuine worth, and these exercises are no exception to the rule. The addition of the logarithm and mensuration questions add greatly to the value."

EDUCATIONAL TIMES:—"The questions have been selected from a great variety of sources: London University Matriculation; Oxford Locals—Junior and Senior; Cambridge Locals—Junior and Senior; Army Preliminary Examinations, etc. As a preparation for examination the book will be found of the utmost value."

A TEXT BOOK OF EUCLID'S ELEMENTS. Including Alternate Proofs, together with additional Theorems and Exercises, classified and arranged. By H. S. HALL, M.A., and F. H. STEVENS, M.A., Masters of the Military and Engineering Side, Clifton College. Globe 8vo. Book I., 1s.; Books I. and II., 1s. 6d.; Books I.-IV., 3s.; Books III. and IV., 2s.; Books III.-VI., 3s.; Books V., VI., and IX., 2s. 6d.; Books I.-VI. and XI., 4s. 6d.; Book XI., 1s.

A KEY *is in preparation.*

CAMBRIDGE REVIEW:—". . . The whole is so evidently the work of practical teachers, that we feel sure it must soon displace every other Euclid."

JOURNAL OF EDUCATION:—"The most complete introduction to Plane Geometry based on Euclid's Elements that we have yet seen."

PRACTICAL TEACHER:—"One of the most attractive books on Geometry that has yet fallen into our hands."

IRISH TEACHERS' JOURNAL:—"It must rank as one of the very best editions of Euclid in the language."

MACMILLAN AND CO., LONDON.

Messrs. Macmillan and Co.'s Mathematical Books.

WORKS BY J. B. LOCK, M.A.

Fourth Edition, Revised. Globe 8vo.

ARITHMETIC FOR SCHOOLS. Complete with Answers, 4s. 6d. Without Answers, 4s. 6d. Part I., with Answers, 2s.; Part II., with Answers, 3s.

KEY. Crown 8vo. 10s. 6d.

Second Edition. Globe 8vo. 2s. 6d.

ARITHMETIC FOR BEGINNERS. A School-Class Book of Commercial Arithmetic.

KEY. By Rev. R. G. WATSON. Crown 8vo. 8s. 6d.

JOURNAL OF EDUCATION:—"This Arithmetic is founded on the author's larger work, such alterations having been made as were necessary to render it suitable for less advanced students, and in particular for those who are candidates for commercial certificates. The book is well adapted to the requirements of this new class of examinations."

PRACTICAL TEACHER:—"Will be found useful in almost all schools, as it contains enough of theory and abundance of practice. The author is thoroughly master of his subject, he is a successful and experienced teacher, and he has the faculty of literary expression. Few writers of mathematical books combine these qualifications in an equal degree."

NATURE:—"A capital handbook."

18mo, Cloth, 1s. With Answers, 1s. 6d.

A SHILLING CLASS-BOOK OF ARITHMETIC. Adapted for use in Elementary Schools.

GUARDIAN:—"This little book differs from most books of the class in that it is based on scientific principles."

SCHOOLMASTER:—"An excellent book of elementary arithmetic. The explanatory portion is pithy but clear, and the examples plentiful, well graduated and fairly difficult."

Just Published. Globe 8vo.

THE TRIGONOMETRY OF ONE ANGLE.

⁎⁎* An easy introduction to Trigonometry containing all the Trigonometry necessary for the Previous Examination at Cambridge.*

Fourth Edition. Globe 8vo. 2s. 6d.

TRIGONOMETRY FOR BEGINNERS, as far as the Solution of Triangles.

KEY. Crown 8vo. 6s. 6d.

SCHOOLMASTER:—"It is exactly the book to place in the hands of beginners. . . . Science teachers engaged in this particular branch of study will find the book most serviceable, while it will be equally useful to the private student."

MACMILLAN AND CO., LONDON.

Messrs. Macmillan and Co.'s Mathematical Books.

WORKS BY J. B. LOCK, M.A.

Seventh Edition. Globe 8vo. 4s. 6d.

ELEMENTARY TRIGONOMETRY.

KEY. By H. CARR, B.A. Crown 8vo. 8s. 6d.

Mr. E. J. ROUTH, Sc.D., F.R.S., writes:—"It is an able treatise. It takes the difficulties of the subject one at a time, and so leads the young student easily along."

NEW ZEALAND SCHOOLMASTER:—"It is a most *teachable* book. Mr. Lock certainly has made good use of his experience in noting difficulties in the student's way and explaining them away. His definitions are all that can be desired. The chapters on logarithms and the use of mathematical tables are ably written."

Fourth Edition. Globe 8vo. 4s. 6d.

HIGHER TRIGONOMETRY. Both Parts in One Volume.
Globe 8vo. 7s. 6d.

CAMBRIDGE REVIEW:—"Obviously the work of one who has had considerable experience in teaching; it is written very clearly, the statements are definite and the proofs concise, and yet a teacher would not find it necessary to add much in the way of supplementary explanation."

Third Edition. Globe 8vo. 4s. 6d.

ELEMENTARY DYNAMICS.

SATURDAY REVIEW:—"The parts that we have more carefully read we have found to be put with much freshness, and altogether the treatment is such as to make the subject interesting to an intelligent pupil."

ENGINEERING:—"This is beyond all doubt the most satisfactory treatise on Elementary Dynamics that has yet appeared."

Second Edition. Globe 8vo. 4s. 6d.

ELEMENTARY STATICS. A Companion Volume to
"Dynamics for Beginners."

NATURE:—"Many treatises deal with elementary statics, but few can rival in clearness Mr. Lock's work. The subject throughout is treated in the author's best style, and the book can be cordially recommended for the use of beginners."

EDUCATIONAL TIMES:—"It is marked by the same clearness and accuracy as the author's *Dynamics*, and affords further testimony of Mr. Lock's ability both as a teacher and as a writer of text-books."

CAMBRIDGE REVIEW:—"We can, without hesitation, recommend this little work as a text-book for beginners."

EUCLID FOR BEGINNERS. Globe 8vo. [*In the Press.*

MACMILLAN AND CO., LONDON.

Messrs. Macmillan and Co.'s Mathematical Books.

BY ISAAC TODHUNTER, M.A., F.R.S.

EUCLID FOR COLLEGES AND SCHOOLS. 18mo. 3s. 6d.
KEY TO EXERCISES IN EUCLID. Crown 8vo. 6s. 6d.
MENSURATION FOR BEGINNERS. With Examples. 18mo. 2s. 6d.
  KEY. Crown 8vo. 7s. 6d.
ALGEBRA FOR BEGINNERS. With numerous Examples. 18mo. 2s. 6d.
  KEY. Crown 8vo. 6s. 6d.
ALGEBRA FOR THE USE OF COLLEGES AND SCHOOLS. Crown 8vo. 7s. 6d.
  KEY. Crown 8vo. 10s. 6d.
TRIGONOMETRY FOR BEGINNERS. With numerous Examples. 18mo. 2s. 6d.
  KEY. Crown 8vo. 8s. 6d.
PLANE TRIGONOMETRY FOR COLLEGES AND SCHOOLS. Crown 8vo. 5s.
  KEY. Crown 8vo. 10s. 6d.
A TREATISE ON SPHERICAL TRIGONOMETRY FOR THE USE OF COLLEGES AND SCHOOLS. Crown 8vo. 4s. 6d.
MECHANICS FOR BEGINNERS. With numerous Examples. 18mo. 4s. 6d.
  KEY. Crown 8vo. 6s. 6d.
A TREATISE ON THE THEORY OF EQUATIONS. Crown 8vo. 7s. 6d.
A TREATISE ON PLANE CO-ORDINATE GEOMETRY. Crown 8vo. 7s. 6d.
  SOLUTIONS. Crown 8vo. 10s. 6d.
A TREATISE ON THE DIFFERENTIAL CALCULUS. Crown 8vo. 10s. 6d.
  KEY. Crown 8vo. 10s. 6d.
A TREATISE ON THE INTEGRAL CALCULUS. Crown 8vo. 10s. 6d.
  KEY. Crown 8vo. 10s. 6d.
EXAMPLES OF ANALYTICAL GEOMETRY OF THREE DIMENSIONS. Crown 8vo. 4s.
AN ELEMENTARY TREATISE ON LAPLACE'S, LAMÉ'S, AND BESSEL'S FUNCTIONS. Crown 8vo. 10s. 6d.
A TREATISE ON ANALYTICAL STATICS. Edited by J. D. EVERETT, M.A., F.R.S. 5th Edition. Crown 8vo. 10s. 6d.

MACMILLAN AND CO., LONDON.

Messrs. Macmillan and Co.'s Mathematical Books.

BY CHARLES SMITH, M.A.,
Master of Sidney Sussex College, Cambridge.

AN ELEMENTARY TREATISE ON CONIC SECTIONS. 7th Edition. Crown 8vo. 7s. 6d.

ACADEMY:—"The best elementary work on these curves which has come under our notice. A student who has mastered its contents is in a good position for attacking scholarship papers at the Universities."

JOURNAL OF EDUCATION:—"We can hardly recall any mathematical text-book which in neatness, lucidity, and judgment displayed, alike in choice of subjects and of the methods of working, can compare with this. . . . We have no hesitation in recommending it as the book to be put in the hands of the beginner."

NATURE:—"A thoroughly excellent elementary treatise."

KEY. Crown 8vo. 10s. 6d.

AN ELEMENTARY TREATISE ON SOLID GEOMETRY. 2nd Edition. Crown 8vo. 9s. 6d.

ACADEMY:—"The best we can say for this text book is that it is a worthy successor to the *Conics* previously noticed by us. . . . Much credit is due for the freshness of exposition and the skill with which the results are laid before the student."

ELEMENTARY ALGEBRA. 2nd Edition. Globe 8vo. 4s. 6d.

SATURDAY REVIEW:—"One could hardly desire a better beginning on the subject."

NATURE:—"For beginners this work should prove invaluable, and even more advanced students would do well to glance over its pages."

EDUCATIONAL TIMES:—"There is a logical clearness about the expositions and the order of chapters for which both schoolboys and schoolmasters should be, and will be, very grateful."

A TREATISE ON ALGEBRA. 2nd Edition. Crown 8vo. 7s. 6d.

SCOTSMAN:—"The exposition is excellently done, and the work is unusually rich in well-chosen exercises and examples. It may be recommended to teachers as a thorough and serviceable text-book."

KEY. Crown 8vo. 10s. 6d.

MACMILLAN AND CO., LONDON.

Messrs. Macmillan and Co.'s Mathematical Books.

ARITHMETIC.

THE GREAT GIANT ARITHMOS. A most Elementary Arithmetic for Children. By MARY STEADMAN ALDIS. Illustrated. Gl. 8vo. 2s. 6d.

ARMY PRELIMINARY EXAMINATION, SPECIMENS OF PAPERS SET AT THE, 1882-89.—With answers to the Mathematical Questions. Subjects: Arithmetic, Algebra, Euclid, Geometrical Drawing, Geography, French, English Dictation. Crown 8vo. 3s. 6d.

A COURSE OF EASY ARITHMETICAL EXAMPLES FOR BEGINNERS. By J. G. BRADSHAW, B.A., Assistant Master at Clifton College. Globe 8vo. 2s. With Answers, 2s. 6d.

ARITHMETIC IN THEORY AND PRACTICE. By J. BROOKSMITH, M.A. Crown 8vo. 4s. 6d.

ARITHMETIC FOR BEGINNERS. By J. and E. J. BROOKSMITH. Globe 8vo. 1s. 6d.

HELP TO ARITHMETIC. Designed for the use of Schools. By H. CANDLER, Mathematical Master of Uppingham School. 2d Edition. Extra fcap. 8vo. 2s 6d.

RULES AND EXAMPLES IN ARITHMETIC. By the Rev. T. DALTON, M.A., Assistant Master at Eton. New Edition, with Answers. 18mo. 2s. 6d.

HIGHER ARITHMETIC AND ELEMENTARY MENSURATION. By P. GOYEN, Inspector of Schools, Dunedin, New Zealand. Cr. 8vo. 5s.

ARITHMETICAL EXERCISES AND EXAMINATION PAPERS. With an Appendix containing Questions in LOGARITHMS and MENSURATION. By H. S. HALL, M.A., Master of the Military and Engineering Side, Clifton College, and S. R. KNIGHT, B.A. Globe 8vo. 2s. 6d.

EXERCISES IN ARITHMETIC for the use of Schools. Containing more than 7000 original Examples. By SAMUEL PEDLEY. Crown 8vo. 5s. Also in Two Parts, 2s. 6d. each.

Works by Rev. BARNARD SMITH, M.A.,
Late Fellow and Senior Bursar of St. Peter's College, Cambridge.

ARITHMETIC FOR SCHOOLS. Cr. 8vo. 4s. 6d. KEY. Cr. 8vo. 4s. 6d.

EXERCISES IN ARITHMETIC. Crown 8vo. 2s. With Answers, 2s. 6d. Answers separately, 6d.

SCHOOL CLASS-BOOK OF ARITHMETIC. 18mo. 3s. Or separately, in Three Parts, 1s. each. KEYS. Parts I., II., and III., 2s. 6d.

SHILLING BOOK OF ARITHMETIC. 18mo. Or separately, Part I., 2d.; Part II., 3d.; Part III., 7d. Answers, 6d. KEY. 18mo. 4s. 6d.

THE SAME, with Answers. 18mo., cloth. 1s. 6d.

EXAMINATION PAPERS IN ARITHMETIC. 18mo. 1s. 6d. The Same, with Answers. 18mo. 2s. Answers, 6d. KEY. 18mo. 4s. 6d.

THE METRIC SYSTEM OF ARITHMETIC, ITS PRINCIPLES AND APPLICATIONS, with Numerous Examples. 18mo. 3d.

A CHART OF THE METRIC SYSTEM, on a Sheet, 42 in. by 34 in. on Roller. 3s. 6d. Also a Small Chart on a Card. Price 1d.

MACMILLAN AND CO., LONDON.

Messrs. Macmillan and Co.'s Mathematical Books.

ALGEBRA.

RULES AND EXAMPLES IN ALGEBRA. By Rev. T. DALTON, Assistant Master at Eton. Part I. 18mo. 2s. *KEY.* Crown 8vo. 7s. 6d Part II. 18mo. 2s. 6d.

ALGEBRAICAL EXERCISES. Progressively Arranged. By Rev. C. A. JONES and C. H. CHEYNE, M.A., late Mathematical Masters at Westminster School. 18mo. 2s. 6d.

KEY. By Rev. W. FAILES, M.A., Mathematical Master at Westminster School. Crown 8vo. 7s. 6d.

ARITHMETIC AND ALGEBRA, in their Principles and Application; with numerous systematically arranged Examples taken from the Cambridge Examination Papers, with especial reference to the Ordinary Examination for the B.A. Degree. By Rev. BARNARD SMITH, M.A. New Edition, carefully revised. Crown 8vo. 10s. 6d.

Works by H. S. HALL, M.A., Master of the Military and Engineering Side, Clifton College, and S. R. KNIGHT, B.A.

ELEMENTARY ALGEBRA FOR SCHOOLS. Sixth Edition, revised and corrected. Globe 8vo, bound in maroon coloured cloth, 3s. 6d.; with Answers, bound in green coloured cloth, 4s. 6d. *KEY.* 8s. 6d.

ALGEBRAICAL EXERCISES AND EXAMINATION PAPERS. To accompany ELEMENTARY ALGEBRA. Second edition, revised. Globe 8vo. 2s. 6d.

HIGHER ALGEBRA. Third edition. Crown 8vo. 7s. 6d. *KEY.* Crown 8vo. 10s. 6d.

GEOMETRICAL DRAWING.

CONSTRUCTIVE GEOMETRY OF PLANE CURVES. By T. H. EAGLES, M.A., Instructor in Geometrical Drawing and Lecturer in Architecture at the Royal Indian Engineering College, Cooper's Hill. Crown 8vo. 12s.

NOTE-BOOK ON PRACTICAL SOLID OR DESCRIPTIVE GEOMETRY. Containing Problems with help for Solutions. By J. H. EDGAR and G. S. PRITCHARD. Fourth edition, revised by A. MEEZE. Globe 8vo. 4s. 6d.

A GEOMETRICAL NOTE-BOOK. Containing Easy Problems in Geometrical Drawing preparatory to the study of Geometry. For the Use of Schools. By F. E. KITCHENER, M.A., Headmaster of the Newcastle-under-Lyme High School. 4to. 2s.

ELEMENTS OF DESCRIPTIVE GEOMETRY. By J. B. MILLAR, Civil Engineer, Lecturer on Engineering in the Victoria University, Manchester. Second edition. Crown 8vo. 6s.

PRACTICAL PLANE AND DESCRIPTIVE GEOMTERY. By E. C. PLANT. Globe 8vo. *[In preparation.*

MENSURATION.

ELEMENTARY MENSURATION. With Exercises on the Mensuration of Plane and Solid Figures. By F. H. STEVENS, M.A. Globe 8vo.
[In preparation.

ELEMENTARY MENSURATION FOR SCHOOLS. By S. TEBAY. Extra fcap. 8vo. 3s. 6d.

MACMILLAN AND CO., LONDON.

Messrs. Macmillan and Co.'s Mathematical Books.

STANDARD BOOKS IN GEOMETRY.

GEOMETRICAL EXERCISES FOR BEGINNERS. By SAMUEL CONSTABLE. Crown 8vo. 3s. 6d.

EUCLIDIAN GEOMETRY. By FRANCIS CUTHBERTSON, M.A., LL.D. Extra fcap. 8vo. 4s. 6d.

A TEXT-BOOK OF EUCLID'S ELEMENTS. Including Alternative Proofs, together with Additional Theorems and Exercises, classified and arranged. By H. S. HALL, M.A., and F. H. STEVENS, M.A., Masters of the Military and Engineering Side, Clifton College. Globe 8vo. Book I., 1s.; Books I. and II., 1s. 6d., Books I.—IV., 3s.; Books V., VI., and XI., 2s. 6d.; Books I.—VI. and XI., 4s. 6d.; Book XI., 1s.
[KEY. In preparation.

THE ELEMENTS OF GEOMETRY. By G. B. HALSTED, Professor of Pure and Applied Mathematics in the University of Texas. 8vo. 12s. 6d.

EUCLID FOR BEGINNERS. Being an Introduction to existing Text-Books. By Rev. J. B. LOCK, M.A. *[In preparation.*

THE PROGRESSIVE EUCLID. With Notes, Exercises, and Deductions. Edited by A. T. RICHARDSON, M.A., Senior Mathematical Master at the Isle of Wight College. Books I. and II. Illustrated. Globe 8vo.
[In the Press.

SYLLABUS OF PLANE GEOMETRY (corresponding to Euclid, Books I.—VI.)—Prepared by the Association for the Improvement of Geometrical Teaching. Crown 8vo. 1s.

SYLLABUS OF MODERN PLANE GEOMETRY. Prepared by the Association for the Improvement of Geometrical Teaching. Crown 8vo. Sewed. 1s.

THE ELEMENTS OF EUCLID. By I. TODHUNTER, F.R.S. 18mo. 3s. 6d. Books I. and II. 1s. KEY. Crown 8vo. 6s. 6d.

Works by Rev. J. M. WILSON, M.A.,
Late Headmaster of Clifton College.

ELEMENTARY GEOMETRY. Books I.—V. Containing the Subjects of Euclid's first Six Books. Following the Syllabus of the Geometrical Association. Extra fcap. 8vo. 4s. 6d.

SOLID GEOMETRY AND CONIC SECTIONS. With Appendices on Transversals and Harmonic Division. Extra fcap. 8vo. 3s. 6d.

Works by CHARLES L. DODGSON, M.A.,
Student and late Mathematical Lecturer, Christ Church, Oxford.

EUCLID. 6th Edition, with Words substituted for the Algebraical Symbols used in the 1st Edition. Books I. and II. Crown 8vo. 2s.

EUCLID AND HIS MODERN RIVALS. 2nd Edition. Crown 8vo. 6s.

CURIOSA MATHEMATICA. A New Theory of Parallels. 3rd Edition. Part I. Crown 8vo. 2s.

MACMILLAN AND CO., LONDON.

Messrs. Macmillan and Co.'s Mathematical Books.

TRIGONOMETRY.

AN ELEMENTARY TREATISE ON PLANE TRIGONOMETRY. With Examples. By R. D. BEASLEY, M.A. 9th Edition, revised and enlarged. Crown 8vo. 3s. 6d.

FOUR-FIGURE MATHEMATICAL TABLES. Comprising Logarithmic and Trigonometrical Tables, and Tables of Squares, Square Roots, and Reciprocals. By J. T. BOTTOMLEY, Lecturer in Natural Philosophy in the University of Glasgow. 8vo. 2s. 6d.

THE ALGEBRA OF CO-PLANAR VECTORS AND TRIGONOMETRY. By R. B. HAYWARD, M.A., F.R.S., Assistant Master at Harrow. *[In preparation.*

A TREATISE ON TRIGONOMETRY. By W. E. JOHNSON, M.A., late Scholar and Assistant Mathematical Lecturer at King's College, Cambridge. Crown 8vo. 8s. 6d.

ELEMENTS OF TRIGONOMETRY. By RAWDON LEVETT and CHARLES DAVISON, Assistant Masters at King Edward's School, Birmingham.
[In the Press.

A TREATISE ON SPHERICAL TRIGONOMETRY. With applications to Spherical Geometry and numerous Examples. By W. J. M'CLELLAND, M.A., Principal of the Incorporated Society's School, Santry, Dublin, and T. PRESTON, M.A. Crown 8vo. 8s. 6d.; or,— Part I. To the End of Solution of Triangles, 4s. 6d.; Part II., 5s.

MANUAL OF LOGARITHMS. By G. F. MATTHEWS, B.A. 8vo. 5s. net.

TEXT-BOOK OF PRACTICAL LOGARITHMS AND TRIGONOMETRY. By J. H. PALMER, Headmaster, R.N., H.M.S. *Cambridge*, Devonport. Globe 8vo. 4s. 6d.

THE ELEMENTS OF PLANE AND SPHERICAL TRIGONOMETRY. By J. C. SNOWBALL, 14th Edition. Crown 8vo. 7s. 6d.

EXAMPLES FOR PRACTICE IN THE USE OF SEVEN-FIGURE LOGARITHMS. By JOSEPH WOLSTENHOLME, D.Sc., late Professor of Mathematics in the Royal Engineering College, Cooper's Hill. 8vo. 5s.

MACMILLAN AND CO., LONDON.

Messrs. Macmillan and Co.'s Mathematical Books.

MECHANICS: STATICS, DYNAMICS.

ELEMENTARY APPLIED MECHANICS. By Prof. T. ALEXANDER and A. W. THOMSON. Part II. Transverse Stress. Crown 8vo. 10s. 6d.

EXPERIMENTAL MECHANICS. A Course of Lectures delivered at the Royal College of Science for Ireland. By Sir R. S. BALL., F.R.S. 2nd Edition, Illustrated. Crown 8vo.

THE ELEMENTS OF DYNAMICS. An Introduction to the Study of Motion and Rest in Solid and Fluid Bodies. By W. K. CLIFFORD. Part I.—Kinematics. Crown 8vo. Books I.-III., 7s. 6d.; Book IV. and Appendix, 6s.

APPLIED MECHANICS: An Elementary General Introduction to the Theory of Structures and Machines. By J. H. COTTERILL, F.R.S., Professor of Applied Mechanics in the Royal Naval College, Greenwich. 8vo. 18s.

LESSONS IN APPLIED MECHANICS. By Prof. J. H. COTTERILL and J. H. SLADE. Fcap. 8vo. 5s. 6d.

DYNAMICS, SYLLABUS OF ELEMENTARY. Part I. Linear Dynamics. With an Appendix on the meanings of the Symbols in Physical Equations. Prepared by the Association for the Improvement of Geometrical Teaching. 4to. 1s.

HYDROSTATICS. By A. G. GREENHILL, Professor of Mathematics to the Senior Class of Artillery Officers, Woolwich. Crown 8vo. [*In preparation.*

ELEMENTARY DYNAMICS OF PARTICLES AND SOLIDS. By W. M. HICKS, Principal and Professor of Mathematics and Physics, Firth College, Sheffield. Crown 8vo. 6s. 6d.

A TREATISE ON THE THEORY OF FRICTION. By JOHN H. JELLETT, B.B., late Provost of Trinity College, Dublin. 8vo. 8s. 6d.

THE MECHANICS OF MACHINERY. A. B. W. KENNEDY, F.R.S. Illustrated. Crown 8vo. 12s. 6d.

KINEMATICS AND DYNAMICS. An Elementary Treatise. By J. G. MACGREGOR, D.Sc., Munro Professor of Physics in Dalhousie College, Halifax, Nova Scotia. Illustrated. Crown 8vo. 10s. 6d.

AN ELEMENTARY TREATISE ON MECHANICS. By S. PARKINSON, D.D., F.R.S., late Tutor and Prælector of St. John's College, Cambridge. 6th Edition, revised. Crown 8vo. 9s. 6d.

LESSONS ON RIGID DYNAMICS. By Rev. G. PIRIE, M.A., Professor of Mathematics in the University of Aberdeen. Crown 8vo. 6s.

Works by JOHN GREAVES, M.A.,
Fellow and Mathematical Lecturer at Christ's College, Cambridge.

STATICS FOR BEGINNERS. Globe 8vo. 3s. 6d.

A TREATISE ON ELEMENTARY STATICS. 2d Ed. Cr. 8vo. 6s. 6d.

HYDROSTATICS FOR BEGINNERS. By F. W. SANDERSON, M.A., Assistant Master of Dulwich College. Globe 8vo. 4s. 6d.

A TREATISE ON DYNAMICS OF A PARTICLE. By Professor TAIT, M.A., and W. J. STEELE, B.A. 6th Edition, revised. Crown 8vo. 12s.

Works by EDWARD JOHN ROUTH, D.Sc., LL.D., F.R.S.,
Hon. Fellow of St. Peter's College, Cambridge.

A TREATISE ON THE DYNAMICS OF THE SYSTEM OF RIGID BODIES. With numerous Examples. Fourth and enlarged Edition. Two Vols. 8vo. Vol. I.—Elementary Parts. 14s. Vol. II.—The Advanced Parts. 14s.

STABILITY OF A GIVEN STATE OF MOTION, PARTICULARLY STEADY MOTION. Adams Prize Essay for 1877. 8vo. 8s. 6d.

MACMILLAN AND CO., LONDON.

Messrs. Macmillan and Co.'s Mathematical Books.

PROBLEMS AND QUESTIONS IN MATHEMATICS.

ARMY PRELIMINARY EXAMINATION, 1882-1889, Specimens of Papers set at the. With Answers to the Mathematical Questions. Subjects: Arithmetic, Algebra, Euclid, Geometrical Drawing, Geography, French, English Dictation. Crown 8vo. 3s. 6d.

CAMBRIDGE SENATE-HOUSE PROBLEMS AND RIDERS, WITH SOLUTIONS:—

1875—PROBLEMS AND RIDERS. By A. G. GREENHILL, F.R.S. Crown 8vo. 8s. 6d.

1878—SOLUTIONS OF SENATE-HOUSE PROBLEMS. By the Mathematical Moderators and Examiners. Edited by J. W. L. GLAISHER, F.R.S., Fellow of Trinity College, Cambridge. 12s.

A COLLECTION OF ELEMENTARY TEST-QUESTIONS IN PURE AND MIXED MATHEMATICS; with Answers and Appendices on Synthetic Division, and on the Solution of Numerical Equations by Horner's Method. By JAMES R. CHRISTIE, F.R.S. Crown 8vo. 8s. 6d.

MATHEMATICAL PAPERS. By W. K. CLIFFORD. Edited by R. TUCKER. With an Introduction by H. J. STEPHEN SMITH, M.A. 8vo. 30s.

SANDHURST MATHEMATICAL PAPERS, for admission into the Royal Military College, 1881-1889. Edited by E. J. BROOKSMITH, B.A., Instructor in Mathematics at the Royal Military Academy, Woolwich. Crown 8vo. 3s. 6d.

WOOLWICH MATHEMATICAL PAPERS, for admission into the Royal Military Academy, Woolwich, 1880-1888 inclusive. Edited by E. J. BROOKSMITH, B.A. Crown 8vo. 6s.

WORKS BY JOSEPH WOLSTENHOLME, D.Sc.,
Late Professor of Mathematics in the Royal Engineering College, Cooper's Hill.

MATHEMATICAL PROBLEMS, on Subjects included in the First and Second Divisions of the Schedule of Subjects for the Cambridge Mathematical Tripos Examination. New Edition, greatly enlarged. 8vo. 18s.

EXAMPLES FOR PRACTICE IN THE USE OF SEVEN-FIGURE LOGARITHMS. 8vo. 5s.

WORKS BY REV. JOHN J. MILNE.

WEEKLY PROBLEM PAPERS. With Notes intended for the use of Students preparing for Mathematical Scholarships, and for Junior Members of the Universities who are reading for Mathematical Honours. Pott 8vo. 4s. 6d.

SOLUTIONS TO WEEKLY PROBLEM PAPERS. Crown 8vo. 10s. 6d.

COMPANION TO WEEKLY PROBLEM PAPERS. Crown 8vo. 10s. 6d.

MACMILLAN AND CO., LONDON.

Messrs. Macmillan and Co.'s Science Class-Books.

Foolscap 8vo.

LESSONS IN ELEMENTARY PHYSICS. By Prof. Balfour Stewart, F.R.S. New Edition. 4s. 6d. (Questions on, 2s.)

EXAMPLES IN PHYSICS. By Prof. D. E. Jones, B.Sc. 3s. 6d.

QUESTIONS AND EXAMPLES ON EXPERIMENTAL PHYSICS: Sound, Light, Heat, Electricity, and Magnetism. By B. Loewy, F.R.A.S. 2s.

A GRADUATED COURSE OF NATURAL SCIENCE FOR ELEMENTARY AND TECHNICAL SCHOOLS AND COLLEGES. Part I. First Year's Course. By the same. Globe 8vo. 2s.

SOUND, ELEMENTARY LESSONS ON. By Dr. W. H. Stone. 3s. 6d.

ELECTRIC LIGHT ARITHMETIC. By R. E. Day, M.A. 2s.

A COLLECTION OF EXAMPLES ON HEAT AND ELECTRICITY. By H. H. Turner. 2s. 6d.

AN ELEMENTARY TREATISE ON STEAM. By Prof. J. Perry, C.E. 4s. 6d.

ELECTRICITY AND MAGNETISM. By Prof. Silvanus Thompson. 4s. 6d.

POPULAR ASTRONOMY. By Sir G. B. Airy, K.C.B., late Astronomer-Royal. 4s. 6d.

ELEMENTARY LESSONS ON ASTRONOMY. By J. N. Lockyer, F.R.S. New Edition. 5s. 6d. (Questions on, 1s. 6d.)

LESSONS IN ELEMENTARY CHEMISTRY. By Sir H. Roscoe, F.R.S. 4s. 6d.—Problems adapted to the same, by Prof. Thorpe. With Key. 2s.

OWENS COLLEGE JUNIOR COURSE OF PRACTICAL CHEMISTRY. By F. Jones. With Preface by Sir H. Roscoe, F.R.S. 2s. 6d.

QUESTIONS ON CHEMISTRY. A Series of Problems and Exercises in Inorganic and Organic Chemistry. By F. Jones. 3s.

OWENS COLLEGE COURSE OF PRACTICAL ORGANIC CHEMISTRY. By Julius B. Cohen, Ph.D. With Preface by Sir H. Roscoe and Prof. Schorlemmer. 2s. 6d.

MACMILLAN AND CO., LONDON.

Messrs. Macmillan and Co.'s Science Class-Books.

Fcap. 8vo, Cloth.

ELEMENTS OF CHEMISTRY. By Prof. IRA REMSEN. 2s. 6d.

EXPERIMENTAL PROOFS OF CHEMICAL THEORY FOR BEGINNERS. By WILLIAM RAMSAY, Ph.D. 2s. 6d.

NUMERICAL TABLES AND CONSTANTS IN ELEMENTARY SCIENCE. By SYDNEY LUPTON, M.A. 2s. 6d.

PHYSICAL GEOGRAPHY, ELEMENTARY LESSONS IN. By ARCHIBALD GEIKIE, F.R.S. 4s. 6d. (Questions on, 1s. 6d.)

ELEMENTARY LESSONS IN PHYSIOLOGY. By T. H. HUXLEY, F.R.S. 4s. 6d. (Questions on, 1s. 6d.)

LESSONS IN ELEMENTARY ANATOMY. By ST. G. MIVART, F.R.S. 6s. 6d.

LESSONS IN ELEMENTARY BOTANY. By Prof. D. OLIVER, F.R.S. 4s. 6d.

DISEASES OF FIELD AND GARDEN CROPS. By W. G. SMITH. 4s. 6d.

LESSONS IN LOGIC, INDUCTIVE AND DEDUCTIVE. By W. S. JEVONS, LL.D. 3s. 6d.

POLITICAL ECONOMY FOR BEGINNERS. By Mrs. FAWCETT. With Questions. 2s. 6d.

THE ECONOMICS OF INDUSTRY. By Prof. A. MARSHALL and M. P. MARSHALL. 2s. 6d.

ELEMENTARY LESSONS IN THE SCIENCE OF AGRICULTURAL PRACTICE. By Prof. H. TANNER. 3s. 6d.

CLASS-BOOK OF GEOGRAPHY. By C. B. CLARKE, F.R.S. 3s. 6d.; sewed, 3s.

SHORT GEOGRAPHY OF THE BRITISH ISLANDS. By J. R. GREEN and ALICE S. GREEN. With Maps. 3s. 6d.

MACMILLAN AND CO., LONDON.

Messrs. Macmillan and Co.'s Scientific Books.

NATURE SERIES.

Crown 8vo.

THE ORIGIN AND METAMORPHOSES OF INSECTS. By Sir John Lubbock, M.P., F.R.S. With Illustrations. 3s. 6d.

THE TRANSIT OF VENUS. By Prof. G. Forbes. With Illustrations. 3s. 6d.

POLARISATION OF LIGHT. By W. Spottiswoode, LL.D. Illustrated. 3s. 6d.

ON BRITISH WILD FLOWERS CONSIDERED IN RELATION TO INSECTS. By Sir John Lubbock, M.P., F.R.S. Illustrated. 4s. 6d.

FLOWERS, FRUITS, AND LEAVES. By Sir John Lubbock. Illustrated. 4s. 6d.

HOW TO DRAW A STRAIGHT LINE; A Lecture on Linkages. By A. B. Kempe, B.A. Illustrated. 1s. 6d

LIGHT: A Series of Simple, Entertaining, and Useful Experiments. By A. M. Mayer and C. Barnard. Illustrated. 2s. 6d.

SOUND: A Series of Simple, Entertaining, and Inexpensive Experiments. By A. M. Mayer. 3s. 6d.

SEEING AND THINKING. By Prof. W. K. Clifford, F.R.S. Diagrams. 3s. 6d.

ON THE COLOURS OF FLOWERS. By Grant Allen. Illustrated. 3s. 6d.

THE CHEMISTRY OF THE SECONDARY BATTERIES OF PLANTÉ AND FAURE. By J. H. Gladstone and A. Tribe. 2s. 6d.

A CENTURY OF ELECTRICITY. By T. C. Mendenhall. 4s. 6d.

THE SCIENTIFIC EVIDENCES OF ORGANIC EVOLUTION. By George J. Romanes, M.A., LL.D. 2s. 6d.

POPULAR LECTURES AND ADDRESSES. By Sir Wm. Thomson. In 3 vols. Vol. I. Constitution of Matter. Illustrated. 6s.—Vol. II. Navigation.

THE CHEMISTRY OF PHOTOGRAPHY. By Prof. R. Meldola, F.R.S. Illustrated. 6s.

ARE THE EFFECTS OF USE AND DISUSE INHERITED? An Examination of the View held by Spencer and Darwin. By W. Platt Hall. 3s. 6d.

MACMILLAN AND CO., LONDON.

MACMILLAN'S SCIENCE PRIMERS.

UNDER THE JOINT EDITORSHIP OF

PROFS. HUXLEY, ROSCOE, and BALFOUR STEWART.

18mo. Cloth. Illustrated. 1s. each.

INTRODUCTORY. By T. H. HUXLEY, F.R.S.

CHEMISTRY. By Sir HENRY E. ROSCOE, F.R.S. With Questions.

PHYSICS. By Prof. B. STEWART, F.R.S. With Questions.

PHYSICAL GEOGRAPHY. By ARCHIBALD GEIKIE, F.R.S. With Questions.

GEOLOGY. By ARCHIBALD GEIKIE, F.R.S.

PHYSIOLOGY. By Prof. M. FOSTER, M.D., F.R.S.

ASTRONOMY. By J. N. LOCKYER, F.R.S.

BOTANY. By Sir J. D. HOOKER, K.C.S.I., F.R.S.

LOGIC. By W. STANLEY JEVONS, F.R.S.

POLITICAL ECONOMY. By W. STANLEY JEVONS, F.R.S.

*** *Others to follow.*

www.ingramcontent.com/pod-product-compliance
Lightning Source LLC
Chambersburg PA
CBHW032246080426
42735CB00008B/1025